Fundamentals of Nano- and Microscale Heat Transport

Arvind Pattamatta · Sarit K. Das

Fundamentals of Nano- and Microscale Heat Transport

Second Edition

Ane Books
Pvt. Ltd.

Arvind Pattamatta
Department of Mechanical Engineering
Indian Institute of Technology Madras
Chennai, Tamil Nadu, India

Sarit K. Das
Department of Mechanical Engineering
Indian Institute of Technology Madras
Chennai, Tamil Nadu, India

ISBN 978-3-031-89612-5   ISBN 978-3-031-89613-2 (eBook)
https://doi.org/10.1007/978-3-031-89613-2

Jointly published with Ane Books Pvt. Ltd.
In addition to this printed edition, there is a local printed edition of this work available via Ane Books in
South Asia (India, Pakistan, Sri Lanka, Bangladesh, Nepal and Bhutan) and Africa (all countries in the
African subcontinent).
ISBN of the Co-Publisher's edition: 978-81-19662-43-2

This Springer imprint is published by the registered company Springer Nature Switzerland AG
The registered company address is: Gewerbestrasse 11, 6330 Cham, Switzerland

If disposing of this product, please recycle the paper.

To Our Families

Preface

It is with great pleasure that we introduce this new book, *Fundamentals of Nano- and Microscale Heat Transport*. In an era where nanotechnology and microscale engineering are advancing at an unprecedented pace, understanding and harnessing heat transport at these scales is of utmost importance. This book fills a crucial gap in the existing literature by providing a comprehensive overview of both nanoscale and microscale energy transport, bringing together these two closely related fields into a single, unified text.

Recent years have witnessed remarkable advancements in the understanding and manipulation of heat transfer phenomena at the nanoscale and microscale. Researchers around the world have been tirelessly exploring the intricacies of heat conduction, convection, and radiation at these dimensions, unlocking novel insights and applications. These advancements hold immense potential for various disciplines, ranging from materials science and engineering to electronics, renewable energy, and biomedical technologies.

Until now, the available textbooks have primarily focused on either nanoscale energy transport or microscale heat transfer as separate entities. This book, however, represents a pioneering effort to bridge the gap and present a holistic approach to studying heat transport phenomena across both scales. By synthesizing the latest research and developments, it equips students, researchers, and practitioners with a comprehensive understanding of the fundamental principles, mathematical models, and experimental techniques pertinent to this rapidly evolving field.

The authors, with their extensive expertise and research contributions in the field of nanoscale and microscale heat transport, have decided to bring out this cohesive text for disseminating knowledge to a wide audience based on several years of teaching Graduate level course on the same topic at the Indian Institute of Technology Madras.

This textbook will serve as an essential companion for graduate-level courses on nanoscale and microscale heat transport. Its well-structured chapters, supplemented with illustrations, and exercise problems, provide an ideal platform for students to grasp the intricacies of the subject and develop a solid foundation. Moreover, this book will prove indispensable for practicing researchers, who will find in it a comprehensive reference guide to deepen their understanding and explore new avenues of inquiry.

The realization of this book would not have been possible without the support of various individuals and institutions. We express our deepest gratitude to the publisher, Ane Books, for recognizing the significance of this work and facilitating its publication. The partnership with IIT Madras, an institution renowned for its excellence in research and education, has been instrumental in ensuring the highest standards of scholarship.

Furthermore, we extend our heartfelt appreciation to the research scholars who have contributed their insights and expertise to this endeavor. Special thanks to V. Siva Sai and Supratim Saha for their help in typesetting the content and in preparing the figures. Their dedication and intellectual contributions have greatly enriched the content of this book. Lastly, we owe a debt of gratitude to our families, whose unwavering support and understanding have allowed us to devote countless hours to the completion of this project.

It is our sincere hope that *Fundamentals of Nano- and Microscale Heat Transport* will inspire readers to delve deeper into this fascinating field May this book serve as a guiding light for students, researchers, and practitioners, facilitating their journey in the field of nanoscale and microscale heat transport.

Chennai, India   Arvind Pattamatta
July 2023   Sarit K. Das

Competing Interests The authors have no competing interests to declare that are relevant to the content of this manuscript.

Contents

About the Authors

Arvind Pattamatta is a professor in the Department of Mechanical Engineering at Indian Institute of Technology Madras. He received his Ph.D. from SUNY, Buffalo, NY in 2009. He is the recipient of Alexander von Humboldt fellowship for the year 2013, INAE Young Engineer Award for the year 2015 and JSPS invitational fellowship for the year 2017. His research interests are in the area micro and nanoscale energy transport. He has authored more than 120 publications in peer-reviewed journals and conferences.

Sarit K. Das is currently the V. Balakrishnan chair professor and the former dean, Academic Research at Indian Institute of Technology Madras. He is also an "institute professor" at IIT Madras. He is a former director of Indian Institute of Technology Ropar. He has published five books and more than 350 research papers. He is an associate editor of *Journal of Heat Transfer Engineering*. His research interests include heat transfer in nano-fluids, biomicro-fluidics, nano-particle-mediated drug delivery in cancer cells, heat exchangers, boiling in mini/micro channels, fuel cells, battery thermal management, jet instabilities, thermal and electrochemical desalination, heat transfer in porous media, and computational fluid dynamics. He is the recipient of DAAD and Alexander von Humboldt Fellowship of Germany.

Symbols

Notations

a	Width of the quantum well, m
A	Cross-sectional area for conduction and surface area for convection, m^2
b	Width of the potential barrier, m
B	A function of the moment of inertia I
Bo	Bond number
Bo	Boiling number
Co	Convection number
C	Specific heat, $Jm^{-3}K^{-1}$
Ca	Capillary number
C_v	Specific heat at constant volume
C_p	Specific heat at constant pressure
d	Diameter and film thickness
D	Length of the standing wave, width of the quantum well, Channel diameter, m, density of states per unit volume, m^{-3}
D_h	Hydraulic diameter, m
e	Charge per electron, spectral density
E	Energy, J
E_c	Conduction band edge, J
E_f	Fermi level, J
Eo	Eotvos number
f	Non-equilibrium distribution function, number density function
f_{eq}	Equilibrium distribution function
$f^{(N)}$	N-particle distribution function
F	Interatomic force, N
g	Gravitational acceleration, ms^{-2} and degeneracy
h	Convective heat transfer coefficient, $Wm^{-2}K^{-1}$ and Planck's constant, Js
$\hbar$	$h/2\pi$, Js
H	Hamiltonian, J
I	Moment of inertia, kgm^2, non-dimensional intensity

Ja	Jakob number
J''	Mass flux, $kgm^{-2}s$
J''_c	Current density, amp.m^{-2}
k	Thermal conductivity, $Wm^{-1}K^{-1}$ and magnitude of wave vector, m^{-1}, spring constant, Nm^{-1}, wave number
K	Wave vector
k_B	Boltzmann constant
l	Azimuthal quantum number, thickness
L	Characteristic length, m
m	Mass, kg, and magnetic quantum number
m_w	Molecular weight
m^*	Effective mass, kg
M	Larger mass, kg
Ma	Mach number
n	Principal quantum number, no. of moles, number density, m^{-3}, Degrees of freedom
N	Number of particles in the system
N_A	Avogadro constant, mol^{-1}
Nu	Nusselt number
p	Momentum, $kgmsec^{-1}$, specularity parameter
P_o	Poiseuille number
Pr	Prandtl number
q	Heat flux vector, Wm^{-2}
Q	Heat transfer rate, W
r	Droplet radius, separation between atoms, m, position vector
R_u	Universal gas constant, $JK^{-1}mol^{-1}$
Ra	Rayleigh number
Re	Reynolds number
S	Seebeck coefficient
t	Time, s
T	Temperature, K
u	Internal energy per unit volume, Jm^{-3}
u	Fluid velocity
U	Internal energy, potential energy, J
v	Molecular instantaneous random velocity, $msec^{-1}$
V	Volume, m^3
W	Power output, W, Wronskian function
We	Weber number
X	Martinelli parameter
x, y, z	Cartesian coordinates
Z	Partition function, Ohnesorge number
$\langle \rangle$	Expectation value

Greek Symbols

α	Thermal diffusivity and absorption coefficient
β	Thermal expansion coefficient, K^{-1}, Lagrange multiplier
θ	Polar angle, temperature
γ	Ratio of gravitational force to surface tension force
κ	Wave vector
κ_B	Boltzmann constant, JK^{-1}
λ	Wavelength, mean free path m
ϵ	Permittivity of free space, F/m
ε	Equivalent roughness
μ	Dynamic viscosity, Nsm^{-2} and chemical potential
ν	Kinematic viscosity, m^2sec^{-1}, linear frequency, sec^{-1}
ρ	Density, kgm^{-3}, speed of light, ms^{-1}
σ	Stefan–Boltzmann constant, $Wm^{-2}K^{-4}$ and electrical conductivity, $\Omega^{-1}m^{-1}$
σ_c	Electrical conductivity, $\Omega^{-1}m^{-1}$
τ	Relaxation time, s
τ_{xy}	Shear stress, Nm^{-2}
Π	Pressure ratio
ϕ	Interatomic potential, J and vector wave form, azimuthal angle, wave function, nanoparticle volume fraction
φ	Electric potential
ψ	Wave function
ω	Angular frequency, $radsec^{-1}$, an operator, solid angle
ξ^*	Accommodation coefficient
ξ^*	Function of accommodation coefficient

Subscripts

app	Apparent
c	Conduction, lower
c	Constricted flow
e	Electron, reduced or equivalent
D	Debye
eq	Equilibrium
f	Fermi
h	Higher
i	Incident
le	Laminar equivalent
m	Mean
r	Reflected
fD	Fermi–Dirac

rad	Radiation
rot	Rotational
trans	Translational
vib	Vibrational
w	Wall
∞	Ambient
s	Surface
sat	Saturation

Superscript

$*$ Complex conjugate, effective property

List of Figures

List of Tables

Chapter 1
Introduction

1.1 Background

The understanding of heat transport processes in small dimensions and timescales is imperative when exploring the unlimited potential that nanotechnology has to offer in areas, such as micro-nanoelectronics, MEMS, and NEMS. In recent years, there has been a great deal of interest in this area, and this is reflected by a surge in the number of research publications and books on this subject. The field of microscale and nanoscale heat transport is quite interdisciplinary, requiring fundamental knowledge of quantum mechanics, statistical thermodynamics, energy states in solids, and classical heat transfer. This book addresses the fundamentals of microscale and nanoscale transport in various fields of current interest, such as thermal dissipation from electronic devices, thermoelectric energy conversion devices, and microelectromechanical systems (MEMS) and sensors. This book is written in an easy-to-comprehend manner to cover all the above-mentioned topics without warranting prerequisites from the interested reader. Students from diverse backgrounds such as mechanical, aerospace, and electrical engineering may find it very helpful as a graduate-level course on this subject, while practicing engineers may find this book as a very useful reference.

1.2 Length Scales and Timescales

Knowledge of length scales and timescales is a prerequisite for the proper understanding of microscale and nanoscale energy transport phenomena. Length scales can be converted into an equivalent timescale, e.g., a length scale of a nanometer can be expressed in timescale in terms of picoseconds or femtoseconds. A key role is played by either local thermal nonequilibrium or local nonequilibrium energy transport in both small length scales and timescales.

© The Author(s) 2025
A. Pattamatta and S. K. Das, *Fundamentals of Nano- and Microscale Heat Transport*,
https://doi.org/10.1007/978-3-031-89613-2_1

1.2.1 Length Scales

The sizes of both microelectromechanical systems (MEMS) and blood cells are typically in the order of 1 micron to 1 mm. The size of semiconductor devices is in the order of 1 nm up to 100 nm. To increase the speed of microprocessors, sub-nanometer semiconductor devices should be packed within a given square meter or square millimeter cross section of microprocessor chips based on Moore's law, resulting in dissipation of more heat from microelectronic devices. Although these devices can be fabricated, controlling the heat flux dissipation from semiconductor devices is a challenge because the order of magnitude of the heat flux is very high, which is equivalent to the heat flux dissipation from the sun. The size of DNA is in the order of a nanometer. Therefore, any device dealing with DNA should be in the order of nanometer or sub-nanometer. These are the typical length scales used in the modern times.

Continuum hypothesis is valid only for the length scale of 1 micron or above. Navier–Stokes equations can be used for the analysis of fluid flow transport as well as constitutive relations. For example, heat flux can be expressed in terms of temperature gradient by using Fourier's law. Navier–Stokes equations can be expressed in terms of velocity and energy equation in temperature rather than stresses. However, researchers from the microprocessor industry scaled down their devices at such a rapid pace and limited the size of their devices in nanometers. Thus, researchers wanted to study the heat dissipation mechanisms, but failed to apply Fourier's law. Hence, subcontinuum use of continuum theories emerged although they were present even before nanometer devices were discovered in physics. Particle-based theories such as the Boltzmann transport equation existed almost 100 years ago, but the applications of sub-micron-level devices emerged only in the past 30 years, and nanometer-sized devices appeared only in the last 5–10 years.

Macro-, micro- and nanodevices can be classified not only based on length scales but also by a non-dimensional parameter, known as Knudsen number, which is the ratio of the mean free path to the characteristic length scale. Mean free path is the average distance travelled by the molecules between two subsequent collisions.

Equilibrium continuum theories may not be applicable and nonequilibrium conditions prevail at very high Knudsen numbers. Knudsen number approaching infinity imply that it is an entirely free molecular limit. So, the molecules hardly can see each other, and they see only the boundaries of the domain. The continuum hypothesis is valid only at small Knudsen numbers. If the Knudsen number is less than 0.001, then it is considered to be in the macroscale regime, where the following can be noted:

1. The continuum approximation and thermodynamic equilibrium are valid.
2. No velocity slip and temperature jump can be observed.

However, if it lies between 0.001 and 0.1 ($0.001 < kn < 0.1$), then the concept of slip applies. Continuum assumption is still valid in the slip flow regime, but the thermodynamic equilibrium will fail. If the Knudsen number is greater than 0.1 and lesser than 10 ($0.1 < kn < 10$), it is called the transition flow. Researchers have been trying to use the higher order formulation of Navier–Stokes equations called extended Navier–Stokes equations for the analysis. However, strictly speaking, extended Navier–Stokes equations may not give the satisfactory results because continuum approximation is not valid in this regime. If the Knudsen number is greater than 10, it is the free molecular flow or rarefied gas flow. Boltzmann transport equation will be used for the analysis.

One has to understand the concept of slip present in the fluid and the boundary. According to the classical theory, there is always a no-slip boundary condition, and the fluid should possess the same velocity as that of the solid. However, with the increasing Knudsen number, slip between fluid and wall becomes relevant. If the Knudsen number is reasonably low, then the continuum hypothesis is still valid, but the Navier–Stokes equations should be modified by correcting few terms in the momentum equation by introducing the finite value of slip at the wall. Therefore, instead of giving a no-slip boundary condition, a partial slip or a slip boundary condition should be considered to solve a particular fluid flow problem. The amount of slip should be obtained from the experimental data.

The Boltzmann transport equation is a subcontinuum model that can also be used in continuum level. The continuum equation can be derived from the Boltzmann equation. The Boltzmann transport equation has two terms: advection term and collision term. Collision between the molecules can still be significant at small Knudsen numbers. As the Knudsen number increases to a tremendous value, no collision is possible between the energy carriers, and the equation becomes a simple hyperbolic equation having only the advection term, which is called the collisionless Boltzmann equation.

The Boltzmann equation can be used to solve for all ranges of Knudsen number. However, the computational effort required to solve the Boltzmann equation is tremendously larger compared to the Navier–Stokes equations, especially at small Knudsen numbers. Hence, for small Knudsen numbers, Navier-Stokes equations are chosen rather than Boltzmann transport equation.

Once the mean free path is known, Knudsen number can be defined based on the mean free path and then different flow regimes can be classified based on the Knudsen number. Let us consider a microchannel with a liquid medium. It is in the continuum range because the mean free path is smaller. However, for microchannels or micropumps with gas flow, the Knudsen number can exceed 0.1. Hence, it can be in the slip or transition flow regime. Therefore, in this case, continuum model Navier–Stokes equations can still be used to solve, but slip boundary conditions are taken into account. At very high Knudsen numbers, continuum approximation is not valid. Hence, subcontinuum models are used.

In recent times, microchannels are gaining much more importance because of the ever-increasing cooling demand for electronic devices. Heat transfer performance in microchannels is improving significantly because of the very low hydraulic diameters, which can be explained by the definition of the Nusselt number. In the case of a circular pipe, for a fully developed laminar flow with constant wall heat flux condition, the Nusselt number is constant and is equal to 4.36, whereas for constant wall temperature boundary condition, it is 3.66. Since the Nusselt number is a function of heat transfer coefficient and hydraulic diameter, with the decreasing dimension of the channel, the heat transfer coefficient increases significantly. Most microchannels are proposed to be used with liquids because gases have low thermal conductivity, and therefore, heat transfer performance is lower.

In microelectronic devices, heat dissipation from the chips to the ambient takes place in three levels. In level 1, heat transfer takes place from the chip, which consists of billions of semiconductor devices, to a directly connected chip carrier. These semiconductor devices are at the nano level. Hence, heat generation takes place and then dissipates to the chip carrier or substrate due to the contact resistance. The chip is bonded to the substrate. Therefore, the heat has to overcome the contact resistance between the chip and the substrate, which is of the order of millimeter square. Hence, it is purely a material phenomenon. Level 2 involves the thermal path from the chip carrier to a casing. Finally, heat rejection takes place from the casing to the ambient. Therefore, in semiconductor devices, microchannel cooling system can be implemented by etching the microchannels into the substrate for effective heat transfer enhancement. However, many times, the processor fails because of the problems with level 1 or level 2. This classification was done by Nakayama in 1988, and it is broadly used in the electronic cooling industry to understand the heat removal process. Hence, this microchannel cooling system can fit into level 2 of cooling, whereas level 1 is purely a materials science problem. More efficient materials with high thermal conductivity and less interfacial thermal resistance should be used for better heat transfer rates.

The macroscopic laws fail under two conditions:

1. If the mean free path of the energy carrier is much higher than the characteristic length scale, then it is considered to be free molecular flow. The continuum hypothesis is not valid in this regime.
2. If the timescales are very less compared to the relaxation time, it is considered to be in the subcontinuum regime, where local nonequilibrium conditions prevail.

Figure 1.1a shows the classical continuum approximation where the mean free path is much smaller than the characteristic length scale between the two plates. The energy carriers are in large numbers, and they have sufficient collisions with each other and also collisions with the boundary of the system.

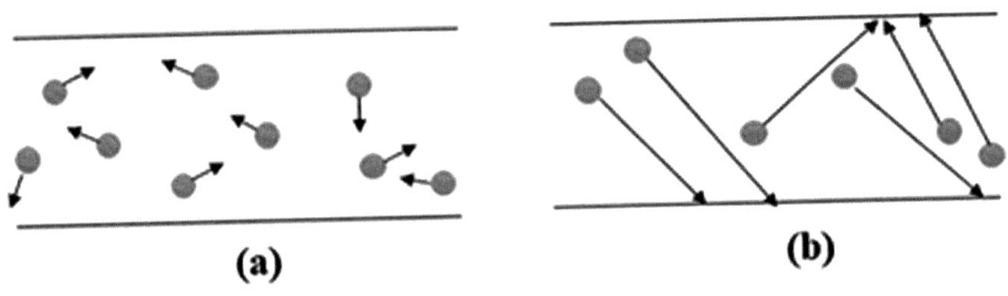

Fig. 1.1 Two regimes: **a** Fourier's law is valid ($\Lambda << L$); **b** Fourier's law is invalid ($\Lambda >> L$), Λ is the mean free path of the heat carrier (phonons or electrons)

On the contrary, in Fig. 1.1b, the mean free path is much larger than the separation between the plates. Therefore, the chances of the intermolecular collisions will be much lesser compared to the collisions between the energy carriers on the boundaries of the system. In this case, one can observe a new phenomenon because of the most predominant collision of the energy carriers with the boundaries, which is attributed to a subcontinuum phenomenon.

1.2.2 Timescales

The timescale is another crucial criterion used to decide whether the continuum approximation is valid. Let us understand the concepts of physical and relaxation timescales with an example. If we are irradiating a laser on a particular material with a pulse width equal to nanoseconds, picoseconds and femtoseconds, then it is called the physical length scale. The average time required for the energy carriers to colloid, come back and relax to an equilibrium state is called the relaxation time. For example, the relaxation time for the electrons in a metal is 10–12 s. At the same time, on irradiating with a pulse width of 10–15 s, then the physical timescale is much smaller than the relaxation timescale. In this case, the continuum laws cannot be used. Hence, to differentiate between continuum and sub-continuum regimes, not only the length scales are essential, but also the timescales are important.

Therefore, one should also understand that anything concerned with subcontinuum is challenging to observe with our sensors and the sensors cannot be used conventionally. For example, temperature can be defined only in the continuum sense. Therefore, a thermocouple cannot be used to measure the temperature at the nano level because the definition of temperature itself is not valid at such small scales. At the nano level, temperature is an indicator of average energy density, but it is not the thermodynamic equilibrium definition of temperature. Therefore, subcontinuum transport equations based on the Boltzmann equation and Monte Carlo simulations are used for analyzing the nanoscale problems.

Figure 1.2 shows the classification of the various levels or regimes of transport on two scales. On the X-axis, the inverse of the Knudsen number is plotted. Continuum regime starts as one moves toward the right-hand side of the X-axis. On the Y-axis, the ratio of the characteristic length to the wavelength is plotted. If we consider the length scale of the order of angstrom, another regime exists within the subcontinuum range where the wave effects become significant. One can use either the wave approach or the particle approach, but the particle approach has a limitation that it can predict well if the device dimension is much larger than the wavelength.

However, when the device dimensions fall below the wavelength, the wave phenomena will become predominant and particle methods cannot predict the wave nature. Therefore, one should apply quantum mechanics, which describes all these energy carriers in the form of waves. The wave nature must be explicitly understood and solved to describe the subcontinuum transport when the length scales fall below the wavelength of the energy carriers. When the characteristic length falls below

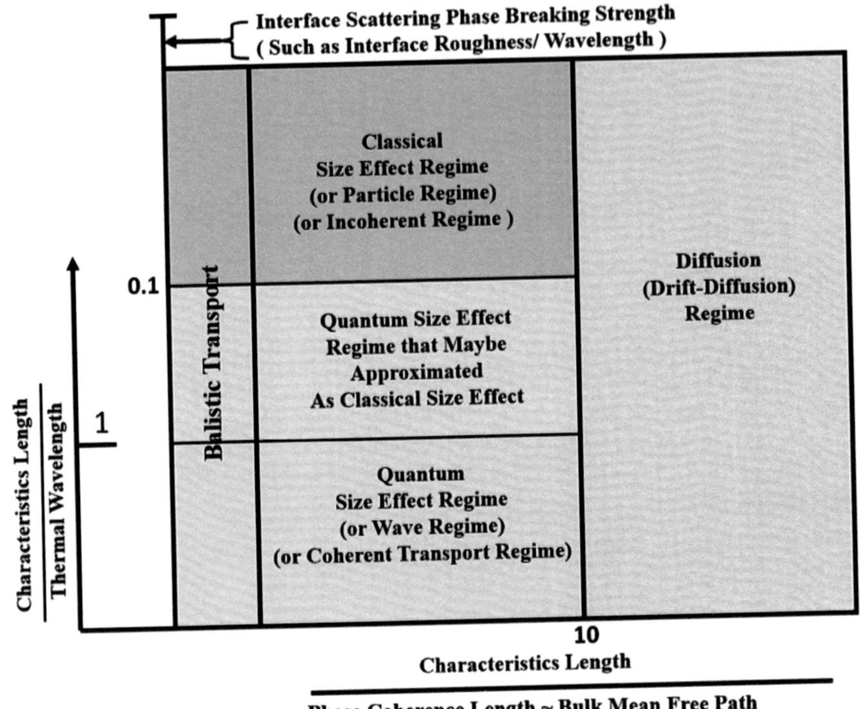

Fig. 1.2 Classification of various regimes for treating phonons as either waves or particles

the wavelength, where the wave regime becomes significant, the Schrodinger wave equation has to be solved. Hence, one cannot explain all the phenomena using only the particle-based method. This means that Boltzmann transport equation alone is not sufficient to describe the entire phenomenon.

Figure 1.3 shows a typical semiconductor device, silicon on insulator (SOI), in which the silicon dioxide substrate, a dielectric, is attached to the top of the silicon film. Here, $Si\,O_2$ is the insulating substrate, over which highly conductive silicon is deposited.

The thickness of the silicon film is in the range of few tens of nanometers, and the silicon oxide substrate might be five times the thickness of the silicon film. A higher temperature is defined at the bottom boundary and a homogenous temperature is defined at the other three boundaries.

The simulation results of a silicon semiconductor device obtained by solving the Boltzmann transport equation are shown in Fig. 1.4. On applying Fourier's approximation and solving for temperature, a set of temperature contours can be seen and no discontinuities in temperature between silicon and SiO_2 are observed. According to the macroscopic laws, the temperature and heat flux should be continuous at the interface. However, when we solve this problem by using a subcontinuum model,

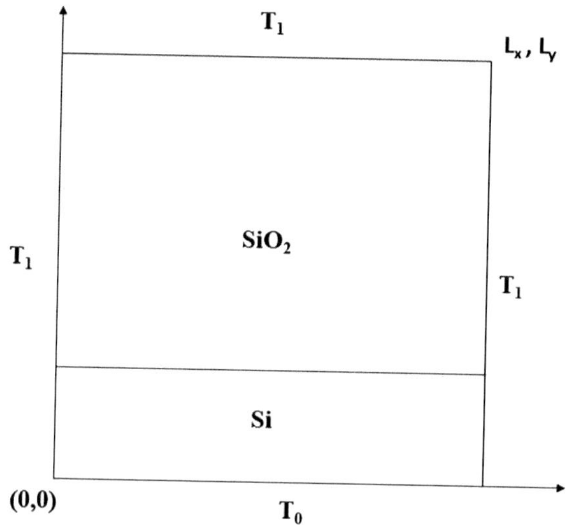

Fig. 1.3 Silicon on insulator (SOI)

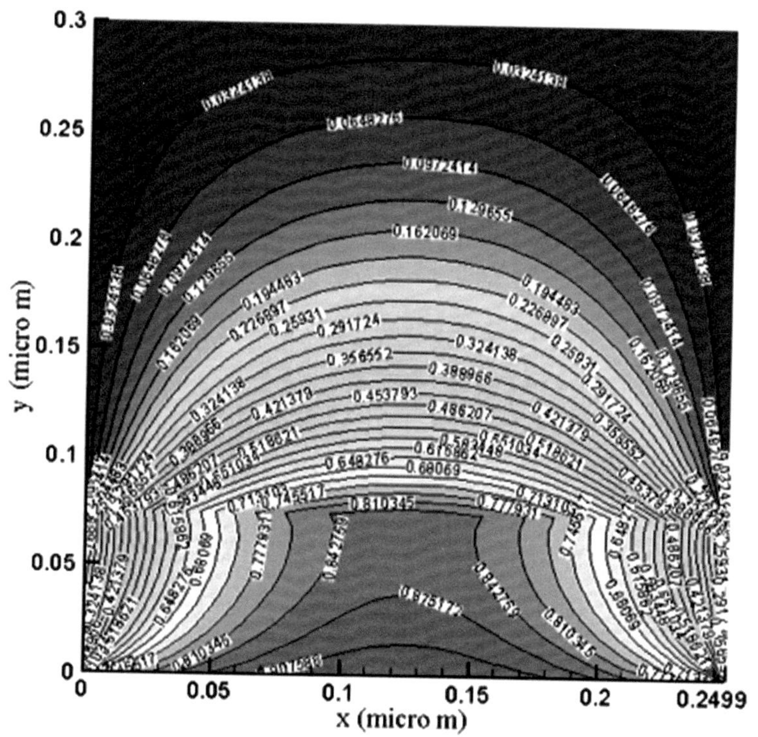

Fig. 1.4 BTE simulation result of a silicon semiconductor device [75]

elliptical contours can be observed in the SiO_2 substrate, which is very close to the solution obtained by Fourier's equation.

However, within the silicon layer, a subcontinuum phenomenon exists and the temperature contours are entirely nonlinear. Hence, there exist a linear pattern of temperature contours outside the silicon layer and a nonlinear pattern of temperature contours within the silicon layer. Furthermore, a temperature discontinuity is observed at the interface. This analysis concludes that the Fourier approximation fails to predict nonequilibrium energy transport processes.

1.3 Review of Thermodynamics and Macroscopic Heat Transfer

Before delving into the nanoscale energy transport, let us start with the simple derivations of heat conduction or heat convection equations in macroscale regime.

Suppose that the system is a closed system and let the surface area of this differential control volume be dA_s and the differential volume be dv, as shown in Fig. 1.5. The normal vector of this element is always pointed outward from the surface. A certain amount of heat is flowing into the system and a specific quantity of work is carried out by the system. We know that the heat transferred into the system and the work done by the system are always positive.

Now, let us derive only a simple heat equation without the effect of work. From the first law of thermodynamics, the change in internal energy is equal to the heat entering into the system, which is Q, and for the differential elemental volume, it is dQ. Therefore, we can integrate this for the entire control volume.

Therefore, from the first law,

$$\frac{dU}{dt} = \dot{Q}. \tag{1.1}$$

Fourier's law can also be used to describe the relation between Q and temperature. Therefore, from Fourier's law,

$$\delta \dot{Q} = -k\nabla T. \tag{1.2}$$

Fig. 1.5 Closed system

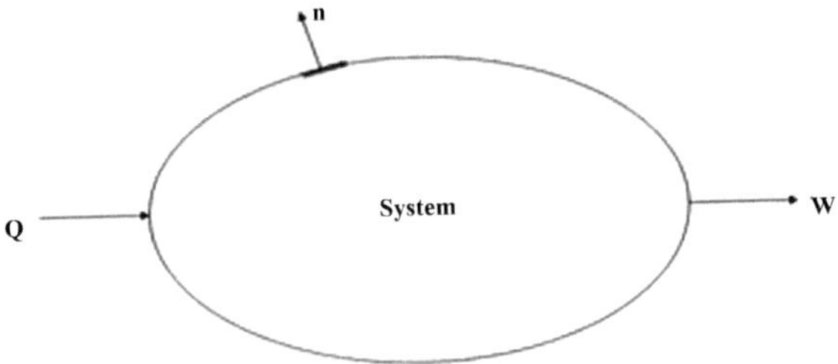

Therefore, for the entire system or control volume, integrating this over the surface area gives the total heat transfer:

$$\int d\dot{Q} = -\int k\nabla T d A_s.$$ (1.3)

The direction of the heat flow and the normal one are opposite to each other, therefore to make this heat transfer positive, another minus sign is inserted.

In terms of specific internal energy,

$$\rho c \int \frac{\partial T}{\partial t} dv = \int k\nabla T d A_s.$$ (1.4)

Applying the Gauss divergence theorem to convert the surface integral to volume integral,

$$\int k\nabla T d A_s = \int \nabla.(k\nabla T) dv.$$ (1.5)

This is the coordinate-free generic heat conduction equation for a closed system.

This analysis can be extended to an open system in which mass transport due to enthalpy is also considered. Consider the velocity entering into the particular control element boundary is $\vec{u}$ and the corresponding enthalpy entering into this particular control volume is $\vec{u} h_i$ and the enthalpy of the fluid leaving this control volume is $\vec{u} h_e$.

Hence, the net efflux of enthalpy leaving this particular differential control volume is $\vec{u}(h_e - h_i)$, simply $\vec{u} h$.

Therefore, by integrating over the entire surface, the total net efflux of enthalpy can be obtained, which is given by

$$\int_A \vec{u} h d A_s.$$ (1.6)

Applying the first law of thermodynamics to an open system,

$$\frac{\partial}{\partial t} \int u dv = \int k\nabla T d A_s - \int_A \vec{u} h d A_s.$$ (1.7)

The net efflux is leaving the system, so it should be on the right-hand side with a minus sign.

On applying the Gauss divergence theorem to the right-hand side,

$$\rho c \int_v \frac{\partial T}{\partial t} dv = \int_v \nabla.(k\nabla T) dv - \int_v \nabla.(\vec{u} h) dv.$$ (1.8)

Now, the relationship between the enthalpy and temperature can be established as follows:

$$\rho C \frac{\partial T}{\partial t} = \nabla.(k\nabla T) - \nabla.(\vec{u}\, h). \tag{1.9}$$

Let us assume that the substance is incompressible. Therefore, the specific heat capacity at constant pressure and at constant volume is same:

$$c_p = c_v = c. \tag{1.10}$$

Therefore, the above equation can be written in terms of temperature:

$$\frac{\partial T}{\partial t} + \nabla.(\vec{u}\, T) = \alpha \nabla^2 T, \tag{1.11}$$

where $\nabla.(\vec{u}\, T)$ is the advection term and $\alpha \nabla^2 T$ is the diffusion term.

1.3.1 Classical Constitutive Relations

1.3.1.1 Fourier's Law of Heat Conduction

Fourier initially proposed the relation between heat flux and temperature gradient. Fourier's law of heat conduction states that the heat flux is directly proportional to the temperature gradient, and the proportionality constant is equal to the thermal conductivity of that particular material. In mathematical form,

$$\vec{Q}'' = -k\nabla T. \tag{1.12}$$

However, this relationship can also be derived from the kinetic theory or from the Boltzmann transport equation.

1.3.1.2 Newton's Law of Cooling

The convective heat transfer rate can be expressed in terms of temperature by using Newton's law of cooling, which can be written mathematically as

$$\vec{Q}'' = h(T_w - T_\infty). \tag{1.13}$$

1.3.1.3 Stefan Boltzmann's Law

The net radiative heat transport between a black surface and the ambient is equal to

$$Q_{rad} = \sigma_E (T_s^4 - T_\infty^4).$$

(1.14)

1.3.1.4 Planck's Law

Planck's law describes the spectral density of electromagnetic radiation emitted by a black body in thermal equilibrium at a given temperature:

$$e_{b,\lambda} = \frac{c_1}{\lambda^5 (e^{c_2/\lambda T} - 1)}.$$

(1.15)

The classical theory states that no body can exceed the black body distribution intensity. However, it is found to exceed at nanoscales. Several phenomena, including phonon tunneling and photon tunneling, are attributed to this improvement at small scales.

1.3.1.5 Newton's Law of Viscosity

Most commonly used constitutive relation in fluid mechanics is Newton's law of viscosity. It states that for a Newtonian fluid, shear stress is directly proportional to the strain rate, and the constant of proportionality is equal to the dynamic viscosity of the fluid:

$$\tau_{xy} = \mu \frac{\partial u}{\partial y}.$$

(1.16)

1.3.1.6 Fick's Law of Diffusion

The most fundamental constitutive relation in mass transfer is Fick's law of diffusion, which is analogous to Fourier's law in heat transfer. It establishes the relationship between the mass flux and the concentration gradient or species gradient. Heat flux (Q''), temperature gradient ($\frac{dT}{dx}$) and thermal conductivity (k) in Fourier's law are analogous to mass flux (J''), concentration gradient ($\frac{dm}{dx}$) and diffusivity (ρD) in Fick's law of diffusion.

$$J'' = -\rho D \frac{dm}{dx},$$

(1.17)

where ρ is the speed of light and D is the diffusion length.

1.3.1.7 Ohm's Law

Ohm's law, the most fundamental constitutive relation in electrical engineering, states that the current density (J_c'') is directly proportional to the electrostatic potential ($\frac{d\phi}{dx}$) and the proportionality constant is equal to the electrical conductivity (σ_c).

$$J_c'' = -\sigma_c \frac{d\phi}{dx}. \tag{1.18}$$

This is analogous to both Fourier's law and Fick's law of diffusion. Some constitutive relations are applicable to various fields, such as heat transfer, fluid mechanics and electrical engineering.

The similarities or analogies between these constitutive relations are also clearly illustrated. However, at high Knudsen numbers, these constitutive relations cannot be used as such, and they are applicable at small Knudsen numbers, which is the limiting condition under which these constitutive relations can be derived from the Boltzmann transport equation. It can be illustrated with a simple scaling analysis to understand the importance of the ratio of relative volume to surface forces acting on a spherical air droplet of radius r. Suitable examples of the volumetric force and surface force are gravitational force and surface tension force, respectively. Assume that a liquid surrounds the droplet. The surface tension force is balancing the pressure difference across the interface between the air inside the droplet and the outside liquid. Therefore, the ratio of gravitational force to surface tension force is proportional to the square of the radius of the spherical droplet.

Let the ratio of gravitational force to surface tension force be denoted with γ. Therefore,

$$\gamma = \frac{\rho g \frac{4}{3} \pi r^3}{\sigma(2\pi r)} = \frac{2\rho g r^2}{3\sigma}. \tag{1.19}$$

Now, consider the value of radius $r = 1$ m, 1 mm, $\rho = 103$ kg/m^3 and $\sigma = 78 \times 10^{-3}$ N/m. At $r = 1$ m, $\gamma = 8.4 \times 10^4$ and at $r = 1$ mm, $\gamma = 8.4 \times 10^{-2}$.

Therefore, on reducing three orders of magnitude in the size of the droplet, there is a sixfold reduction in the value of γ. Therefore, gravitational force is significant compared to surface tension force for a bubble of 1-m diameter. In the case of channel flow problems also, gravity plays a vital role in flow patterns compared to microchannel and minichannel flows. Hence, specific forces become very significant at different length scales. It is a perfect example where the relative importance of forces becomes significant while analyzing the flow physics of a particular problem.

1.4 Overview of Microscale and Nanoscale Transport Phenomena

To understand the macroscale transport phenomena, one should have a thorough knowledge about microscale nature of energy carriers. Let us consider a box filled with plenty of gas molecules. One side of the box is maintained at a higher temperature and the other side at a lower temperature. This system is still at a macroscale where continuum approximations are valid. Heat is transferred from the hot side to the cold side through conduction. As time progresses, the average kinetic energy of the molecules on the hot side increases. Therefore, an average number of collisions increases on the hot side, and gradually these collusions transfer the energy from the hot side to the cold side. This is called the diffusion process. At the macroscale, the heat flux and the rate of heat transfer can be related to the temperature gradient. However, at the microscale, it is the primarily collision and transfer of energy from one molecule to the other.

In liquids and gases, essentially molecules are responsible for conduction. However, in the case of dielectric solids or semiconductors, lattice vibrations are responsible for heat conduction because no medium exists for the energy transport. These lattice vibrations travel at the speed of sound in the lattice, which is the maximum speed at which they can travel. Directly, these lattice vibrations can be represented as massless imaginary particles called phonons.

An absolute force is acting between two atoms in a lattice structure, which holds the atoms in specific positions. These phonons are mainly described by the statistical distribution function called the Bose–Einstein distribution function.

In metals, conduction band electrons are abundant, which are primarily responsible for heat conduction, as shown in Fig. 1.6. The energy carriers associated with thermal radiation are electromagnetic waves called photons. Thermal radiation is a result of oscillation of the electromagnetic waves, particularly in the bandwidth under solar radiation.

A spring—mass system can represent the interactive potential between two atoms in a lattice structure, as shown in Fig. 1.7. If the atoms come close together, the spring will push them out, whereas if they move away from each other, then the spring will get compressed. This spring—mass system assumption is used to calculate the forces of interaction between the molecules.

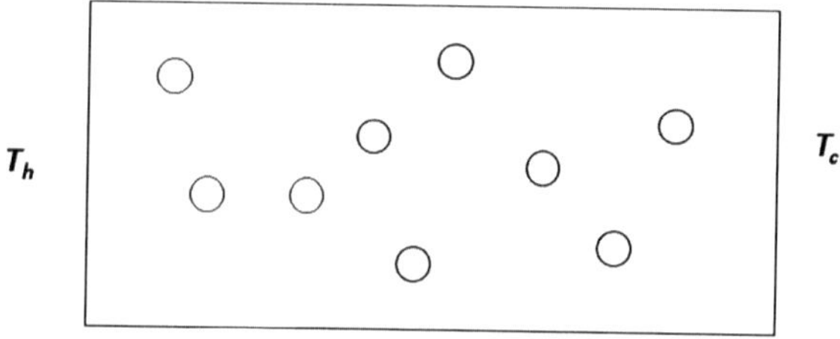

Fig. 1.6 Heat conduction process

Fig. 1.7 Spring mass system

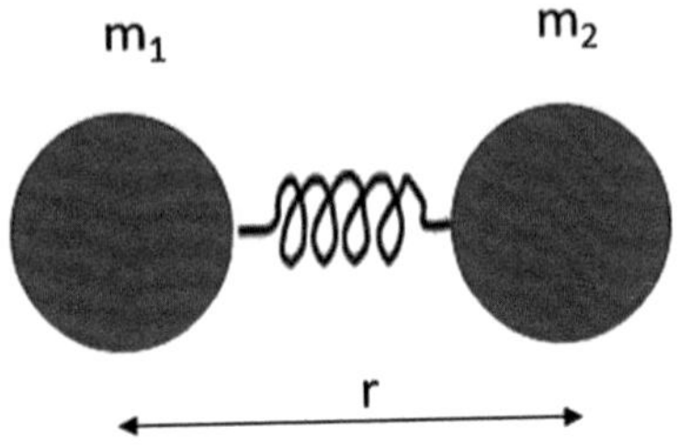

The interactive potential is drawn on the Y-axis and the spacing between the atoms on the X-axis. The potential on the positive Y-axis is positive potential (attractive potential) and that on the negative Y-axis is negative potential (repulsive potential). When the spacing is minimal, an extreme negative repulsive potential acts between the atoms. As the separation between the two atoms increases, the atoms attain stability. When the separation distance increases further, an attractive potential will generate and the curve moves toward the positive potential, and, finally, the curve is asymptotic to the X-axis. This repulsive potential value is equal to $\frac{A}{r^{12}}$. The attractive potential is defined by a curve that is $\frac{B}{r^6}$. The overall potential between the two atoms is written as $\frac{A}{r^{12}} - \frac{B}{r^6}$. This is called the Lennard-Jones potential, a widely used model in molecular dynamic simulations.

Bosons are particles that are primarily described by the Bose–Einstein distribution function. Bosons are indistinguishable, i.e., they all look similar and any number of particles can occupy a given quantum state or energy state. Quantum state is a discrete energy level. These discrete energy levels are the integral multiples of $h\upsilon$. Therefore, according to Planck's law,

$$E = nh\upsilon, \tag{1.20}$$

where $n = 1, 2, 3, \ldots$ and h is the Planck constant and it's value is equal to 6.626×10^{-34} J. s. Photons also come under the category of bosons and are described by the Bose–Einstein distribution.

Electrons are described by the Fermi–Dirac distribution function, and their nature is also indistinguishable. However, for a given spin, electrons will have both plus half and minus half, spin quantum states. Hence, only one electron can occupy a particular spin quantum state. Therefore, electrons obey Pauli's exclusion principle in terms of occupation of the quantum states because of the spin.

Molecules are distinguishable, i.e., characteristics of molecules of a particular gas or a liquid are different from those of another gas or liquid. There is no limit for the number of molecules that can occupy the energy level like bosons. Molecules are governed by a classical Maxwell–Boltzmann distribution function. Hence, the energy levels are not quantized like Bose–Einstein or Fermi–Dirac distribution functions. The derivations of these distribution functions are covered in statistical thermodynamics.

1.4.1 Frequency or Energy Range

The frequency or energy range of molecules, electrons, and photons is 0 to ∞. Electrons can occupy the energy states from 0 to ∞, depending upon temperature and material. Phonons cannot occupy all the energy levels ranging from 0 to ∞. A maximum possible frequency limit exists for phonons to which the energy is limited in the energy states called the Debye cutoff limit.

1.4.2 Order of Magnitude of Velocities

Each of the energy carriers is also limited by the corresponding velocities that they can possess. Therefore, the order of magnitude of velocity is different for different energy carriers. For phonons, the order of magnitude of velocity is equal to the speed of sound in a given lattice. In general, the acoustic sound is of the order of 10^3 m/s. The speed of free electrons is of the order of 10^6 m/s and for the molecules, it should be of the order of 10^2 m/s. The speed of light is of the order of 10^8 m/s, which is the same as that of the order of magnitude of velocity for photons since they travel with the speed of light in vacuum.

1.4.3 Mean Free Path

Mean free path is an essential characteristic to understand the subcontinuum transport phenomenon. The average distance between two phonons in a lattice before they collide is in the range of 10–100 nm. Hence, the mean free path of phonons is on the order of 10–100 nm. Electrons also possess a mean free path similar to that of phonons. For gas molecules, it should be equal to 1 nm, whereas for liquids, it is of the order of 100 nm.

1.4.4 Wavelength

The wavelength of phonons and molecules is of the order of 1 nm and that of electrons is of the order of 10–50 nm. However, for photons, the wavelength is equal to 10 nm–1 km. The wavelength of photons is longer because electromagnetic waves operate in a vacuum, whereas phonons, electrons, and molecules operate in a specific medium.

So far, the energy carriers in different media have been briefly described. A summary of these energy carriers is given in Table 1.1.

The amount of energy or momentum these energy carriers possess should be clearly understood. Therefore, particle nature and wave aspects of these energy car-

Table 1.1 Comparison between the energy carriers based on various attributes

Attributes	Phonons	Electrons	Molecules	Photons
Source	Lattice vibrations	Free from nucleate bonding	Atoms	Electromagnetic vibration or oscillation
Propagation media	Solids, dielectrics, semiconductors, metals	Metals	Liquids or gases	Vacuum
Energy presentation	Wave vector, frequency, modes and polarization	Wave vector	Momentum, kinetic and potential energies	Frequency polarization
Distribution function	Bose–Einstein (bosons)	Fermi–Dirac (fermions)	Maxwell–Boltzmann distribution	Bose–Einstein (bosons)
Nature of particles	Indistinguishable	Indistinguishable	Distinguishable	Indistinguishable
Particle distribution	Any number of particles can occupy a given quantum state	Obeys Pauli's exclusion principle	Any number of particles can occupy a given energy level	Any number of particles can occupy a given quantum state
Equilibrium distribution	$\dfrac{1}{e^{-\frac{\hbar\omega}{k_B T}}-1}$	$\dfrac{1}{e^{\frac{E_e-\mu}{k_B T}}+1}$	$\dfrac{1}{e^{\frac{m(k_x^2+k_y^2+k_z^2)}{2k_B T}}}$	$\dfrac{1}{e^{-\frac{\hbar\omega}{k_B T}}-1}$
Frequency or energy range	Debye cutoff limit	$0-\infty$	$0-\infty$	$0-\infty$
Velocities (m/sec)	$\sim 1e03$	$\sim 1e06$	$\sim 1e02$	$\sim 1e08$
Mean free path	10–$100\,nm$	10–$100\,nm$	4–$100\,nm$	$10\,nm$–$1\,km$
Wave length	$\sim 1\,nm$	~ 10–$50\,nm$	$\sim 1\,nm$	$10\,nm$–$1\,km$

riers can be explained through classical mechanics and quantum mechanics, respectively. Hence, the energy carriers can be classified based on classical mechanics regime and quantum mechanics regime.

Three significant microscopic energies decide the average total energy possessed by the particle, which contributes to the macroscale internal energy. Internal energy in the macroscale is the sum of three microscopic energies: translational, rotational and vibrational energies.

Let the mass of the particle be m kg, moving with a velocity of v m/sec. Then the translational energy is equal to

$$E_{trans} = \frac{mv^2}{2} = \frac{p^2}{2m}, \tag{1.21}$$

where momentum $p = mv$. If we move from particle nature to wave nature, then the momentum is equal to $p = \hbar k$.

Where the modified Plank's constant $\hbar = \frac{h}{2\pi}$ and the wave vector $k = \frac{2\pi}{\lambda}$. Therefore, $p = \frac{h}{\lambda}$ or $\lambda = \frac{h}{p}$ is called the de Broglie wavelength. The magnitude of wave vector k can be expressed as

$$k^2 = k_x^2 + k_y^2 + k_z^2, \tag{1.22}$$

where k_x, k_y and k_z are continuous waves.

In quantum mechanics, the translational energy can be obtained by solving the Schrodinger wave equation:

$$E_{trans} = \frac{\hbar k^2}{2m}. \tag{1.23}$$

The Schrodinger wave equation is a partial differential equation and can be solved by the method of separation of variables. This can be solved analytically by using particle in a box or quantum well approach. A single particle exists inside the quantum well, and it is always confined inside the well because the potential outside the well will be infinite. Within the quantum well, the particle has certain potential energy u. In quantum mechanics, the energy levels are discontinuous, and they get discretized because of quantum confinement.

In classical mechanics, the rotational energy can be obtained by the expression

$$E_{rot} = \frac{I\omega^2}{2}, \tag{1.24}$$

where I and ω are the moment of inertia and angular velocity, respectively. In quantum mechanics, rotational energy can be obtained by solving the Schrodinger equation for the simplest hydrogen atom. The expression for the rotational energy is given by

$$E_{rot} = hBl(l + 1), \tag{1.25}$$

where $l = 0, 1, 2, \ldots$, B is a function of the moment of inertia I and l is an integer. In classical mechanics, the concept of vibrational energy can be very well understood with the spring—mass system. In this system, the atom can be assumed as a mass and the forces of attraction between the atoms can be assumed to have a spring constant-like nature. Damping effects can be neglected. Therefore, the equation of motion for the spring—mass system can be written as

$$\frac{d^2x}{dt^2} + \frac{c}{m}x = 0. \tag{1.26}$$

The solution for this equation is given by

$$x(t) = c_1 \cos \omega t + c_2 \sin \omega t, \tag{1.27}$$

Table 1.2 Comparison between various modes of energy in classical and quantum mechanics

Modes of energy	Classical mechanics	Quantum mechanics
Translational energy	$E_{trans} = \frac{mv^2}{2} = \frac{p^2}{2m} = \frac{\hbar^2 k^2}{2m}$	$E_{trans} = \frac{\hbar^2 k^2}{2m}$
Rotational energy	$E_{rot} = \frac{I\omega^2}{2}$	$E_{rot} = hBl(l+1)$ where $l = 0, 1, 2, \ldots$
Vibrational energy	$E_{vib} = \frac{mv^2}{2} + \frac{kx^2}{2}$	$E_{vib} = hv(n + \frac{1}{2})$

where $\omega = \sqrt{\frac{k}{m}}$.

The boundary conditions for this system are as follows: at $t = 0$, $x = 0$ and $\frac{dx}{dt} = 0$.

The constants C_1 and C_2 can be obtained by substituting the boundary conditions. Therefore, the total vibrational energy is the sum of kinetic energy and potential energy.

$$E_{vib} = KE + PE = \frac{1}{2}mv^2 + \frac{1}{2}kx^2. \tag{1.28}$$

On substituting v and x into this expression,

$$E_{vib} = \frac{1}{2}mc_1^2\omega^2. \tag{1.29}$$

Similarly, quantum mechanical vibrational energy can be obtained by solving the Schrodinger equation. The vibrational energy is given by

$$E = hv(n + {}^1\!/_2). \tag{1.30}$$

where $v = \frac{1}{2\pi}\sqrt{\frac{k}{m}}$ and $n = 0, 1, 2, 3, \ldots$.

The comparison between various modes of energy in classical and quantum mechanics are summarized in Table 1.2.

1.5 Research Applications: Microelectronics and Information Technology

The critical thermal issues in nanoscale transport include the effect of small length and timescales on the energy transport. The generation of hot spots at small length scales can be mitigated by considering SOI devices. In the nanoscale regime, the hot spot mitigation raises the thermal bottleneck problem. Hotspots are generated due to the nonequilibrium between electrons and phonons and the thermal resistance between the metal and the dielectric at microscopic length scales. With Fourier's equation at these small length scales, the hot spot temperature and the heat flux values cannot be predicted accurately. The next aspect is the removal of hot spots. With intense research on many novel materials, researchers have found that carbon nanotube arrays can improve thermal conductivity in a particular direction. Therefore, over the last decade, carbon nanotubes have emerged as the potential material for microelectronics because electrical and thermal conductivities can be tuned to be enormously high in a particular direction. They can be used as thermal interface materials. The hot spot generation can also be observed at small timescales.

In the case of lasers, pulse width ranges from milliseconds to femtoseconds. Therefore, the actual physical timescale for heat to penetrate from the surface to subsurface is much longer compared to that of a radiating pulse. Therefore, there possibly exists strong local nonequilibrium, which leads to hot spot generation.

Therefore, many applications are dealing with nonequilibrium not only in lasers but also in optoelectronics and nanophotonics.

1.5.1 Microfabrication and Nanofabrication

In his visionary talk in 1959, Feynman envisioned the future of controlling and manipulating devices at microscopic scales, for example, writing (with an electron beam) all the 24 volumes of Encyclopedia Britannica on the head of a pin. In 1983, he addressed quantum computing. In the 1990s, micromachining and MEMS devices emerged as an active research area. Later on, they were developed as tools for biological and medical diagnostics such as lab-on-a-chip, with pump valve and analysis sections on the 10- to 100-micron scale. Microchannels and microscale heat pipes have also been developed and tested for electronic cooling applications.

Fabrication is another major issue concerned with microscale and nanoscale devices. Conventional machining tools cannot perform microfabrication and nanofabrication. Besides, microfabrication and nanofabrication techniques are being used in electrical engineering for a long time. Photolithography is a technique used to fabricate microelectronic devices and semiconductor chips in electrical engineering as well as to create mechanical devices. Typical MEMS devices or the lab-on-chip devices can also be fabricated using photolithography. However, mechanical cutting is currently used rather than photolithography.

Figure 1.8 shows the dust mite that cannot be seen with the naked eye. The size of dust mites are in microns. Some gears manufactured by MEMS technology can be compared to the size of a dust mite.

Therefore, researchers from chemistry and materials science have been working for the last 30–40 years to develop the new processes for synthesizing materials at the nanoscale.

Fig. 1.8 Dust mite

1.5.2   *Probing of Small Structures*

Characterization is another essential aspect after the fabrication of nanostructures, such as carbon nanotubes and nanoparticles. Conventional microscopes cannot be used for the characterization of nanoscale structures. Hence, techniques based on electron microscopy should be used.

1.5.2.1   Transmission Electron Microscope (TEM)

A TEM uses thermal excitation or applies a high voltage to draw electrons from the tip end. The electrons are then accelerated by an intense electrical field to gain a significant momentum p ($\lambda \sim 1$ angstrom). Since the resolution is usually comparable to wavelength λ, high resolution is obtained with electron energy as high as MeV magnitude. The electrons penetrate through the sample (less than 200 nm thick) and the diffraction or transmission can be observed from the detector.

1.5.2.2   Scanning Electron Microscope (SEM)

Different from TEM, SEM only observes the surface and electrons do not penetrate into the sample. Electrons have lower energy in the SEM. TEM images can be seen in Fig. 1.9.

The resolution of the TEM is quite high, which is in the order of angstroms, whereas the SEM can only resolve surface aberrations, which are of the order of few

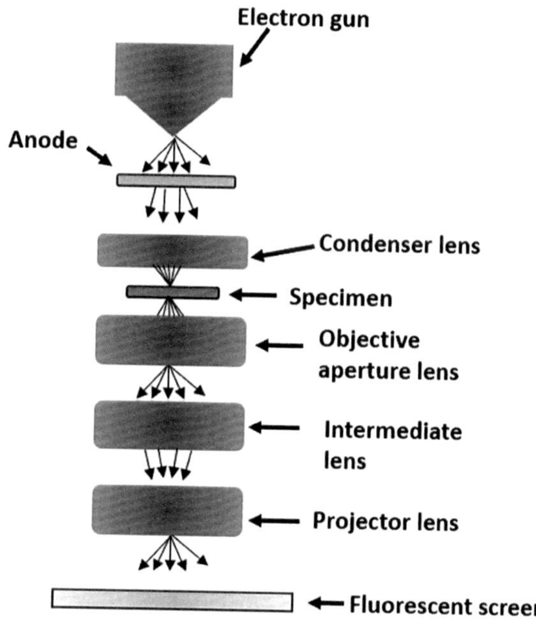

Fig. 1.9 Transmission electron microscope

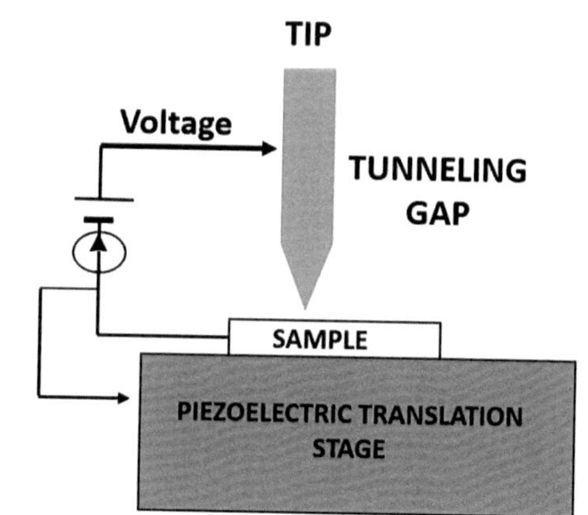

Fig. 1.10 Scanning tunneling electron microscope

nanometers. Therefore, depending on the resolution level, either an SEM or a TEM can be used.

Another most important instrument used to probe the nanostructures is the scanning tunneling microscope (STM) (see Fig. 1.10, and it is based on the tunneling phenomena). The tunneling phenomena form the basis for several inventions that led to several Nobel prizes, including the tunneling diode by Esaki (1958) and the scanning tunneling electron microscope (STM) Binnig and Rhorer, 1982). In STM, a sharp tip is brought in close proximity to a conducting surface but not contacting the surface. The piezoelectric stage can adjust the distance between the tip and the

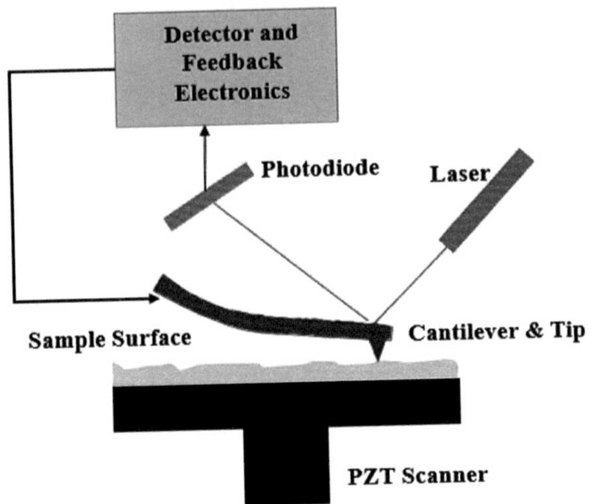

Fig. 1.11 Atomic force microscope

sample with subatomic accuracy. When voltage is applied, electrons tunnel through the vacuum gap and create a current in the loop. The electrons directly appear, on the other hand, rather than flowing from the tip to the sample. This phenomenon is called tunneling.

STM cannot be used to scan dielectric surfaces because they need to be conductive to provide tunneling electrons. For nonconductive surfaces, AFM was invented in Stanford University. The basic operation of AFM is shown in Fig. 1.11. A diamond is attached to an Al film to form a scanning probe. STM is mounted on top of the Al film to detect its deflection. Since the spring constant between atoms is much larger than that of the beam (Al film), while scanning the beam will bend according to the surface topography, the atoms will not be affected. In this way, the surface image is obtained. In the current AFM, the STM is replaced with a laser beam.

1.5.3 Thermoelectric Energy Conversion

Thermoelectricity refers to a class of phenomena in which the temperature difference creates an electric potential or an electric potential creates a temperature difference. The Seebeck effect and Peltier effect are used for power generation and refrigeration, respectively.

1.5.3.1 The Seebeck Effect

When two junctions are maintained at two different temperatures, an electric potential will be induced. This phenomenon is called the Seebeck effect. The hot junction acts as a heat source and the cold junction acts as a heat sink. Some external work is being done while transferring heat from the hot side to the cold side. Electric potential is generated due to the transfer of electrons and holes from the source to the sink in the n-type and p-type semiconductors, respectively.

1.5.3.2 The Peltier Effect

The Peltier effect is the reverse of the Seebeck effect. In the Peltier effect, when a current is passed between two junctions, heat can be absorbed and deposited on either side, respectively—this is the principle used in the refrigeration process.

Thermoelectricity arises from the difference between the average energy of conduction electrons and Fermi energy. From the solid-state physics, the density of states describes the average density of electrons that can occupy at each energy level. The actual occupational probability is given by the Fermi–Dirac distribution function, which comes from the statistical thermodynamics. The pictorial representation of the density of states and Fermi–Dirac distribution function is shown in Fig. 1.12. Fermi–Dirac distribution function is an equilibrium distribution function, which governs the probability of occupational distribution of electrons.

The density of occupied states is the product of the density of states and Fermi–Dirac distribution function. The density of occupied states shows that most of the average energy is focused on the conduction band. This average energy is the average energy of the conduction electrons. As mentioned earlier, this difference between the average energy of conduction electrons and Fermi energy gives rise to thermoelectricity.

Therefore, the higher the gap between these two, the more amount of current is generated for a fixed temperature difference. Similarly, for a given current, a temperature difference can be obtained. Therefore, the energy difference between the conduction electrons and the Fermi level is the driving potential for thermoelectricity.

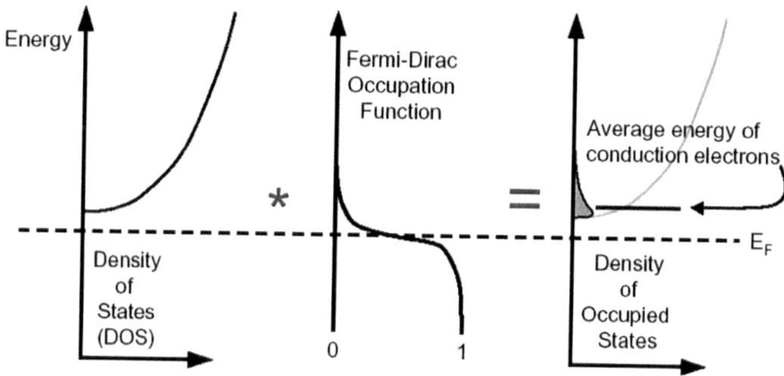

Fig. 1.12 Pictorial representation of density of states and Fermi–Dirac distribution function [75]

1.5.4 Material Selection

The band gap between the conduction band and Fermi level is larger for insulators, moderate for semiconductors and smaller for metals. For insulators, the total number of electrons in the conduction band is negligible. Therefore, the average energy of the conduction electrons is quite small.

On the other hand, in metals, there exists the tremendous energy of electrons in the conduction band, but the band gap between the conduction band and the Fermi level is almost 0. Therefore, a quantity called power factor is introduced, which is an indicator of thermoelectricity. It is the product of S^2 and the electrical conductivity (σ), where S is the Seebeck coefficient.

$$ZT = \frac{S^2 \sigma}{k} T. \tag{1.31}$$

This means that the higher the difference between the conduction band and the Fermi energy, the higher the value of the Seebeck coefficient. In the case of insulators, electrical conductivity is small, but the value of S is quite significant. Therefore, the product of $S^2\sigma$ becomes considerably small. For metals, the sigma is very large and S is quite small. Therefore, the power factor value is minimal. In the case of semiconductors, optimally high values of $S^2\sigma$ can be observed due to moderately large values of S and σ.

Therefore, superior materials for thermoelectric power generation are semiconductors and semiconductor-based materials can be used for producing thermoelectricity.

1.5.5 Dimensionless Figure of Merit

To non-dimensionalize the power factor $S^2\sigma$, it should be divided by thermal conductivity (k) and multiplied with temperature (T). This is called the non-dimensional figure of merit.

As the power factor increases, thermal conductivity also increases. The desirable conditions to have larger thermoelectricity are as follows: the temperature difference must be maintained between the hot and cold sides, and the electrons and holes only should flow between the hot and cold sides. To meet these conditions, low thermal conductivity materials are required. If thermal conductivity is very low, heat conduction from the hot to the cold side reduces. If the heat also flows between the hot and cold plates, then the amount of thermoelectricity naturally reduces. Therefore, the dimensionless figure of merit brings low thermal conductivity value and high power factor, thereby rendering excellent device performance.

1.5.6 Thermoelectric Power Generation: Applications

Thermoelectric power generation has an extensive range of applications. For example, the Voyager spacecraft cannot be powered by solar panels. Therefore, thermoelectricity is used to power these spacecraft. In addition, some of the navy electrical ships also use thermoelectricity. Thermoelectricity can be used in waste heat recovery of cars and power plants. There are some portable thermoelectric refrigerators that cool the fluid at 5° for the extended period.

Chapter 2
Fundamentals of Quantum Mechanics

2.1 Introduction

A fundamental knowledge of quantum mechanics is a prerequisite to understand the wave effects at the nano level, since wave effects are dominant in nanoscale energy transport. This fact can be elucidated with a simple example where quantum effects are predominant in the nanoscale regime. Consider a thin film of semiconductor material made of germanium or silicon. The thickness (L) of the thin film could be of the order of 1–100 nm. One end of the thin film is maintained at a higher temperature (T_1) and the other end at a lower temperature (T_2). Heat transfer takes place from the hot side to the cold side by conduction. The thermal conductivity of the thin film can be calculated by using Fourier's law of heat conduction:

$$Q = kA \frac{T_1 - T_2}{L}.$$

(2.1)

This is the classical method for measuring thermal conductivity. Several transient methods are also available for calculating thermal conductivity in which the temperature cannot be measured directly, but the pseudo-temperature can be measured.

One of the most commonly used transient methods is the 3-omega technique. In this method, a voltage is passed through a thin film at a frequency ω, and then the third harmonic or three times omega component of this voltage is extracted. The transient heat conduction equation should be plugged into this system. The third harmonic is a direct function of the increase of temperature across the thin film, which, in turn, is a function of thermal conductivity. Therefore, plotting thermal conductivity as a function of the thickness of the thin film (l), at $l = 1000$ nm, it is observed that the value of k is close to the bulk value because this is the thickness at which microscale or nanoscale effects can be neglected. Therefore, Fourier's heat conduction equation should be used to estimate the thermal conductivity in this continuum regime. For example, for silicon, the mean free path is of the order of 100 nm. Therefore, the Knudsen number is less than or equal to 0.1. Hence, the

© The Author(s) 2025

A. Pattamatta and S. K. Das, *Fundamentals of Nano- and Microscale Heat Transport*,
https://doi.org/10.1007/978-3-031-89613-2_2

value of k approaches the macroscale thermal conductivity value. In general, the thermal conductivity value of bulk materials is constant. However, in the 3-omega technique, thermal conductivity decreases with the thickness of the thin film. If the film thickness is in the range from 10 nm to a few hundreds of nanometers, the nanoscale effects become dominant. Therefore, thermal conductivity decreases with the film thickness in this region. As the size becomes smaller than the mean free path, there are chances that energy carriers can collide with the boundaries rather than with each other. Therefore, diffusion is negligible compared to the scattering at the boundaries. Hence, boundary scattering is the most dominant mechanism at nanoscales. Therefore, a considerable jump in temperature occurs at the boundaries. This temperature can be sensed only through a diffusion process. Since only very few energy carriers are left at the boundary, they directly come and hit the boundary, which results in an artificial temperature drop. Owing to this temperature drop, thermal conductivity is reduced in this regime.

As the film thickness decreases to less than 10 nm, wave effects start to emerge. As a result, quantum and mechanical effects appear. Most importantly, a gradual increase in thermal conductivity can be observed in this regime. The enhancement in the value of thermal conductivity is caused by the transmission of phonons across the boundaries. This phenomenon is called phonon tunneling and can be observed very clearly at the interface of two materials. For example, in a semiconductor device made of germanium and silicon, phonons can be easily tunneled across the interface from one material to the other. As a result, the thermal conductivity increases as the thickness of the thin film decreases below 10 nm.

2.2 Basic Wave Characteristics

Waves can be classified into two types: traveling waves and nontraveling waves. For a traveling wave, the wave function could be a sinusoidal or a cosine distribution function. If one assumes a size distribution function, then the wave function ϕ can be written as a *sine* distribution function in both time and space:

$$\phi = \sin(\omega t - kx), \tag{2.2}$$

where the angular frequency $\omega = 2\pi\upsilon$, where υ is the linear frequency and the wave number $k = \frac{2\pi}{\lambda}$, and the angular frequency (ω) and wave number (k) represent the wave characteristics over time and space, respectively. The wave function (ϕ) can be plotted against both time and space. In the ϕ versus t graph (Fig. 2.1), the distance between two successive peaks is called the time period, which is the reciprocal of frequency:

$$T = \frac{1}{\upsilon}. \tag{2.3}$$

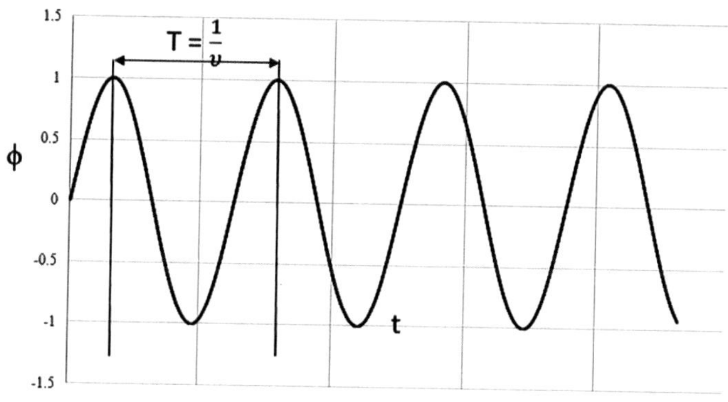

Fig. 2.1 The ϕ versus t graph

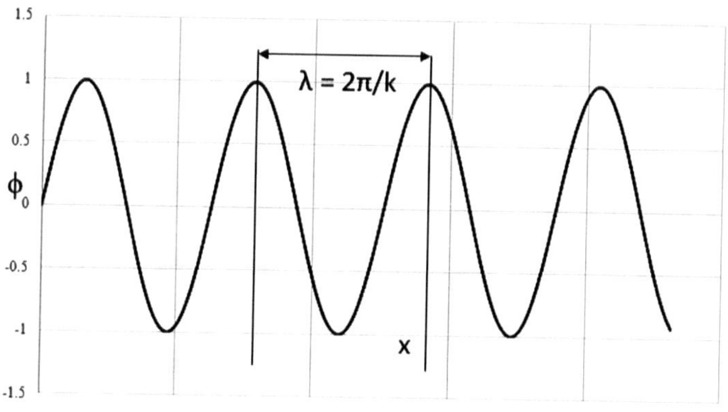

Fig. 2.2 The ϕ versus x graph

In the ϕ versus x graph (Fig. 2.2), the distance between two successive peaks is called the wavelength (λ):

$$\lambda = \frac{2\pi}{k}.$$

(2.4)

The complex representation of the wave is given by

$$Ae^{-i(\omega t - kx)} = A[\cos(\omega t - kx) - i\sin(\omega t - kx)].$$

(2.5)

For a wave with a fixed phase, $\omega t - kx = $ constant. On differentiating with respect to x, $\frac{dx}{dt} = \frac{\omega}{k}$ is called the phase velocity, which represents the speed of propagation of the wave. Therefore, on plotting waves with a fixed phase ($\omega t - kx = $ constant), inclined lines having a positive slope can be obtained for each $\frac{\omega}{k}$. These are called right-running waves. Similarly, left-running waves can be obtained from the equation $\omega t + kx = $ a constant. A linear superposition of these two waves constitutes the standing wave, as shown in Fig. 2.3.

Fig. 2.3 A standing wave

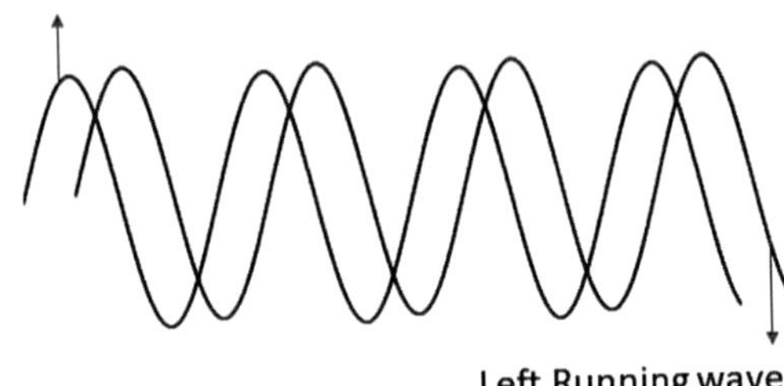

Fig. 2.4 A standing wave
pinned at both ends

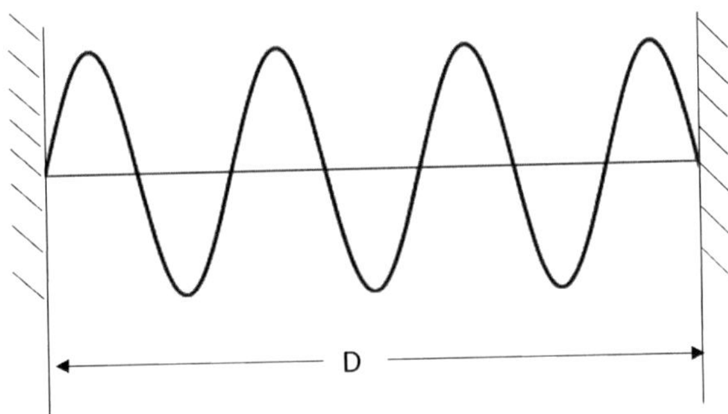

In complex coordinates, the right-running and left-running waves can be represented
by e^{ikx} and e^{-ikx}, respectively. On superposing these two waves, a standing wave
can be obtained:

$$\phi = Ae^{ikx} + Be^{-ikx}. \tag{2.6}$$

The standing wave appears to be pinned at both ends, as shown in Fig. 2.4. Hence, it
cannot move either right or left. Therefore, the typical values of wave function will
be 0 at the boundaries. At $x = 0, D$, $\phi = 0$, where D is the length of the standing
wave.

The standing wave can be represented as

$$\phi = A \cos kx + B \sin kx. \tag{2.7}$$

On applying the boundary conditions, the following can be noted:

(i) At $x = 0$, $\phi = 0$, we obtain $A = 0$. Therefore, $\phi = B \sin kx$.
(ii) At $x = D$, $\phi = 0$. Therefore, $B \sin kD = 0 \Rightarrow k = \frac{n\pi}{D}$.

Here, $n = 1, 2, 3, \ldots, \infty$. Therefore, $k = \frac{n\pi}{D}$. On considering the particle nature, the
energy of these waves is given by

$$E = \frac{p^2}{2m}, \tag{2.8}$$

where p is the particle momentum. According to quantum mechanics, the momentum can be written as

$$p = \hbar k. \tag{2.9}$$

Therefore, upon substituting p into the expression for energy, one can obtain

$$E = \frac{p^2}{2m} = \frac{(\hbar k)^2}{2m} = \frac{\hbar^2}{8m}\left(\frac{n}{D}\right)^2. \tag{2.10}$$

On varying the value of n, different modes of these waves can be represented diagrammatically.

If $n = 1$, then $k = \frac{\pi}{D}$. Hence, only one peak can be observed.

If $n = 2$, then $k = \frac{2\pi}{D}$. Therefore, one peak and one trough can be observed.

Similarly, n can be equated to $1, 2, 3, \ldots, \infty$. Different wave patterns can be observed for different values of n. Hence, the standing wave within confinement can be discretized into an infinite number of waves of different modes. Therefore, an infinite number of standing waves can be created within the confinement and each of the waves can have a specific amount of energy (Fig. 2.5).

Therefore, the expression for the change in energy (ΔE) between the successive values of n can be written as

$$\Delta E = \frac{\hbar^2}{8m D^2}(n_2^2 - n_1^2). \tag{2.11}$$

In the case of the electron, the mass of the electron (m_e) = 9.1×10^{-31} kg, Planck's constant (h) = 6.6×10^{-34} J sec, and confinement length (D) = 1 mm. Therefore, the change in energy between state 1 ($n = 0$) and state 2 ($n = 1$) is given by $\Delta E = 1.7 \times 10^{-31}$ J.

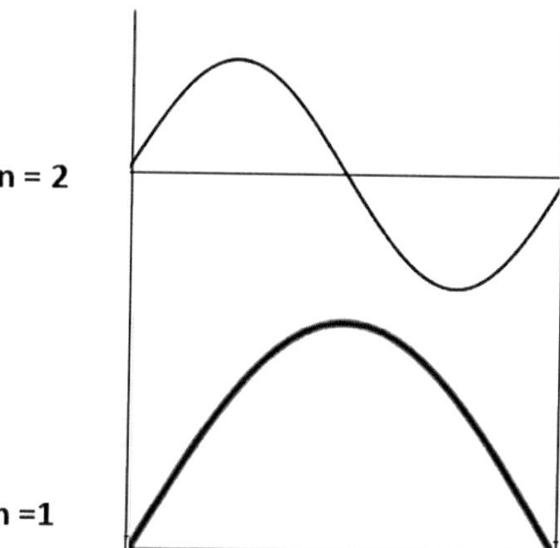

Fig. 2.5 Standing waves of different modes within a confinement

Now, the energy associated with room temperature $k_B T = 4.1 \times 10^{-21} \text{J}$, where k_B is Boltzmann's constant. Therefore, it is clear that the change in energy between two successive levels is less than that of the designable energy at $D = 1$ mm. This means that one cannot even sense the energy shift caused by quantization at room temperature. At $D = 10^{-8}$m, the change in energy between two successive levels $\Delta E = 10^{-21}$J. Hence, the order of ΔE is comparable to $k_B T$. Therefore, if the confinement size is huge, then the shift or spacing between the energy levels is too small. Therefore, the discreteness or quantization can be neglected. On the contrary, if the confinement length is minimal, then the shift becomes very large. Therefore, real quantization happens at the smallest confinement sizes. The confinement length for the waves should be of the order of nm.

2.3   Schrodinger Wave Equation

The most fundamental equation of physics for describing the quantum mechanical behavior of waves is the Schrodinger wave equation. The Schrodinger wave equation is given in a mathematical form as

$$-\frac{\hbar^2}{2m}\nabla^2\psi + U\psi = i\hbar\frac{\partial\psi}{\partial t}. \tag{2.12}$$

The term $-\frac{\hbar^2}{2m}\nabla^2 + U$ is called the Hamiltonian operator. Using the Hamiltonian operator, the Schrodinger equation can be modeled as follows:

$$H\psi = i\hbar\frac{\partial\psi}{\partial t}. \tag{2.13}$$

The Hamiltonian is an indicator of the energy possessed by the wave. Here, ψ is the wave function. It could be a pressure wave or an electric field. The first term in this equation is the Laplacian operator in space and the second term is the potential energy constraint. The term on the right-hand side is the temporal derivative describing the evaluation of the wave function in time.

If the potential energy constraint in this equation is 0, then the wave is free to move without any constraint. Therefore, the equation becomes

$$-\frac{\hbar^2}{2m}\nabla^2\psi = i\hbar\frac{\partial\psi}{\partial t}. \tag{2.14}$$

This is a hyperbolic equation. The imaginary number i makes this wave equation hyperbolic. Interestingly, Schrodinger himself did not find the correct explanation for the wave function ψ. Later on, a scientist named Born concluded that there is a significance to ψ only if it is multiplied with the complex conjugate of itself. For example, if $\psi = a + ib$, its complex conjugate will be given as $\psi^* = a - ib$.

The product $\psi \,.\psi^*$ is the probability density function, which gives the probability of finding a particle at a particular position in space. Therefore, according to the normalization condition,

$$\int_{-\infty}^{\infty} \psi\psi^* = 1. \tag{2.15}$$

In quantum mechanics, there is always an uncertainty associated with estimating the quantities. They are called expected quantities. Primarily, these expected quantities become more probabilistic at tiny scales because the actual value that has to be measured is less than the uncertainty itself.

The expected value of the position of a particle in space is given by

$$\langle r \rangle = \int_{-\infty}^{\infty} \psi r \psi^*, \tag{2.16}$$

where r could be x, y, and z. The expected value of the position of a particle in space in the general form is given by

$$\langle \Omega \rangle = \int_{-\infty}^{\infty} \psi^* \Omega \psi dx, \tag{2.17}$$

where Ω could be $\vec{r}$, time (t), momentum (p), and energy.

The expected value of momentum is given by

$$\Omega = -i\hbar \nabla \vec{r} = -i\hbar \left[\frac{\partial}{\partial x} \hat{x} + \frac{\partial}{\partial y} \hat{y} + \frac{\partial}{\partial z} \hat{z} \right]. \tag{2.18}$$

Similarly, the expected value of energy is given by

$$\Omega = H = \frac{p^2}{2m} + U = \frac{\hbar^2 \nabla^2}{2m} + U. \tag{2.19}$$

Therefore,

$$H\psi = i\hbar \frac{\partial \psi}{\partial t}. \tag{2.20}$$

In general, the standard deviation is an indicator of uncertainty. For example, the standard deviation in a random quantity x can be defined as

$$\Delta x = \sqrt{\frac{1}{n-1}\sum_{i=1}^{n}(x-\langle x\rangle)^2}. \qquad (2.21)$$

In quantum mechanics, the uncertainty in a random quantity Q can be defined as

$$\langle \Delta Q\rangle = \sqrt{\int \psi^*(Q-\langle Q\rangle)^2\psi dx}. \qquad (2.22)$$

Using this expression for uncertainty, the Heisenberg uncertainty principle can be proved directly. This principle states that it is impossible to determine the position and velocity of a particle simultaneously. Therefore,

$$\langle \Delta p\rangle.\langle \Delta x\rangle \geq \frac{\hbar}{2}. \qquad (2.23)$$

The expected uncertainty in energy and time can be defined as

$$\langle \Delta E\rangle.\langle \Delta t\rangle \geq \frac{\hbar}{2}. \qquad (2.24)$$

2.4 Solution to Schrodinger's Wave Equation

The Schrodinger wave equation can be solved by ignoring the potential energy constraint because retaining the potential energy constraint increases the complexity of solving the equation.

This equation can be considered as a partial differential equation with nonconstant coefficients. The solution can be obtained by the separation of variables method, which is the direct method for solving simple partial differential equations:

$$\psi(r,t) = \psi(r).Y(t). \qquad (2.25)$$

The wave function (ψ) is a function of both position (r) and time (t). The solution can be obtained by considering the wave function as the product of two independent solutions.

On differentiating with respect to space and time, Schrodinger's wave equation becomes

$$\frac{1}{\psi}\left[\frac{-\hbar^2}{2m}\nabla^2\psi + U\psi\right] = i\hbar\frac{1}{Y}\frac{dY}{dt}. \qquad (2.26)$$

This equation is valid when both left-hand side and right-hand side terms are equal to a constant. Therefore,

$$\frac{1}{\psi}\left[\frac{-\hbar^2}{2m}\nabla^2\psi + U\psi\right] = i\hbar\frac{1}{Y}\frac{dY}{dt} = E \Rightarrow H\psi = E\psi. \tag{2.27}$$

On solving this equation, the solution in time can be obtained as

$$Y(t) = C_1 e^{\frac{-iEt}{\hbar}}. \tag{2.28}$$

The expected value of the Hamiltonian can be written as

$$\langle H\rangle = E \Rightarrow \langle H\rangle = \int_{-\infty}^{\infty} \psi^* H\psi dx. \tag{2.29}$$

Substituting $\psi = \psi(r)Y(t)$, the expected value of the Hamiltonian becomes

$$\langle H\rangle = \int_{-\infty}^{\infty} \{\psi(r)Y(t)\}^* H\{\psi(r)Y(t)\}dx. \tag{2.30}$$

Rearranging the terms, we obtain

$$\langle H\rangle = \int_{-\infty}^{\infty} \{\psi(r)Y(t)\}^* H\psi(r)\{Y(t)\}dx. \tag{2.31}$$

Substituting $H\psi(r) = E\psi(r)$ and rearranging the terms, we obtain

$$\langle H\rangle = \int_{-\infty}^{\infty} \{\psi(r)Y(t)\}^* E\{\psi(r)Y(t)\}dx. \tag{2.32}$$

Writing E outside the integral, we obtain

$$\langle H\rangle = E\int_{-\infty}^{\infty} \{\psi(r)Y(t)\}^* \{\psi(r)Y(t)\}dx \tag{2.33}$$

$$\Rightarrow \langle H\rangle = E\int_{-\infty}^{\infty} \psi^*\psi dx = E. \tag{2.34}$$

Therefore, the expected value of the Hamiltonian is nothing but energy (E).

2.4.1 Free Particles in 1-D Space

The Schrodinger wave equation in the general form can be written as

$$\frac{-\hbar^2}{2m}\nabla^2\psi + (U - E)\psi = 0. \tag{2.35}$$

Therefore, to solve this equation, the potential energy constraint U should be known. Since this is a second-order ODE, two boundary conditions are required.

The Schrodinger wave equation for a free particle in 1-D space can be written as

$$\frac{-\hbar^2}{2m}\nabla^2\psi = E\psi. \tag{2.36}$$

This is the second-order homogeneous ODE with constant coefficients and the solution can be written as

$$\psi(x) = Ae^{-ikx} + Be^{ikx}, \tag{2.37}$$

where $k = \frac{\sqrt{2mE}}{\hbar}$. Note that the potential energy constraint is zero because a particle is free to go anywhere in 1-D space. Therefore, the final solution in both time and space can be obtained as follows:

$$\psi(x, t) = Ae^{-i(\frac{E}{\hbar}t + kx)} + Be^{-i(\frac{E}{\hbar}t - kx)}. \tag{2.38}$$

Here, the first term corresponds to the left-running wave and the second term corresponds to the right-running wave. Hence, the combination of waves running on the right and left can be presented as the solution to the wave function, and there is also an equal probability of finding these waves running on both right and left sides.

2.4.2 Particles in a 1-D Confinement

A potential well or a quantum well is a region of space where a particle is free to move in a 1-D space, as shown in Fig. 2.6. Therefore, the potential energy constraint of the particle within the well is zero. There exists a confinement outside the well such that the particle cannot move out of the well. Hence, the potential energy outside the well is infinity. This indicates that to move out of the well, the particle should have an infinite amount of energy to overcome the potential energy barrier, which is impossible. Hence, the probability of finding a particle outside the quantum well is zero. In terms of the wave function, $\psi = 0$.

Let the total width of the quantum well be D. The coordinate system can be established at the leftmost corner of the potential well such that the width varies

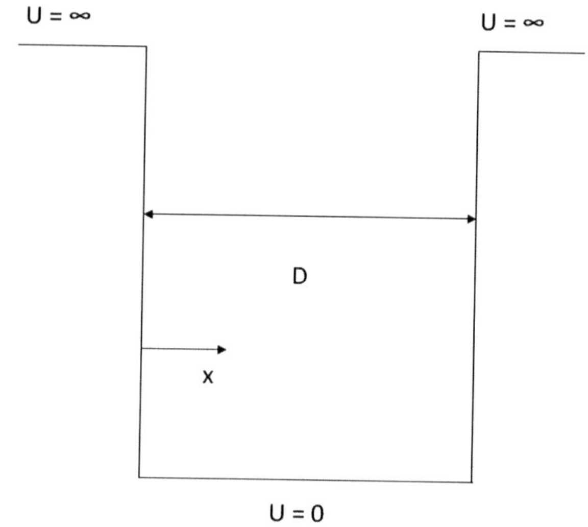

Fig. 2.6 A particle in a 1-D confinement

between $x = 0$ and $x = D$. The width could be in nm. Therefore, the boundary conditions can be written as follows: at $x = 0$, $\psi = 0$ and at $x = D$, $\psi = 0$.

The Schrodinger wave equation inside the potential well can be written as

$$-\frac{\hbar^2}{2m}\frac{\partial^2 \psi}{\partial x^2} - E\psi = 0. \tag{2.39}$$

This can be solved by using the separation of variables method in which the wave function can be written in terms of both space and time. The solution in time could be an exponential function and the solution in space can be obtained by solving an eigenvalue problem. On applying the boundary conditions, the expression for the eigenvalue can be obtained as energy, which is the expected value of the Hamiltonian.

The solution in space can be obtained as

$$\psi(x) = A\cos kx + B\sin kx, \tag{2.40}$$

where $k = \sqrt{\frac{2mE}{\hbar^2}}$.

Applying the boundary condition, $\psi(x = 0) = 0 \Rightarrow kD = n\pi \Rightarrow k = \frac{n\pi}{D}$,

$$E = \frac{1}{2m}\left(\frac{\pi \hbar n}{D}\right)^2. \tag{2.41}$$

Therefore,

$$\psi_n(x) = B\sin\frac{n\pi x}{D}, \tag{2.42}$$

where $n = 1, 2, 3, \ldots, \infty$.

The constant B can be obtained by using the following normalization criterion:

$$\int_{x=-\infty}^{\infty} \psi^* \psi \, dx = 1. \tag{2.43}$$

Therefore,

$$B = i\sqrt{\frac{1}{2D}}. \tag{2.44}$$

The expected value of the position of the particle is given by

$$\langle x \rangle = \int_{0}^{D} \psi^* x \psi \, dx = \frac{D}{2}. \tag{2.45}$$

From this result, one can conclude that the position of the particle is most likely at the centre of the quantum well.

2.4.3   Particle in a 2-D Confinement

In the 1-D potential well, the particle can only move in the positive and negative x-directions. However, in the 2-D quantum well shown in Fig. 2.7, the particle can also move along the y-axis. Inside the well, the potential energy constraint is 0, and at the boundaries of the well, the potential energy becomes ∞. Therefore, the particle cannot escape from the 2-D well.

The Schrodinger wave equation for the 2-D quantum well can be written as

$$-\frac{\hbar^2}{2m}\left[\frac{\partial^2 \psi}{\partial x^2} + \frac{\partial^2 \psi}{\partial y^2}\right] - E\psi = 0. \tag{2.46}$$

The separation of variables approach gives a solution. Therefore, the wave function can be written as

$$\psi(x, y) = X(x)Y(y). \tag{2.47}$$

Dividing the Schrodinger wave equation by $X(x).Y(y)$, we obtain

$$\frac{1}{X}\frac{\partial^2 X}{\partial x^2} + \frac{1}{Y}\frac{\partial^2 Y}{\partial y^2} + \frac{2mE}{\hbar^2} = 0. \tag{2.48}$$

Fig. 2.7 A particle in a 2-D confinement

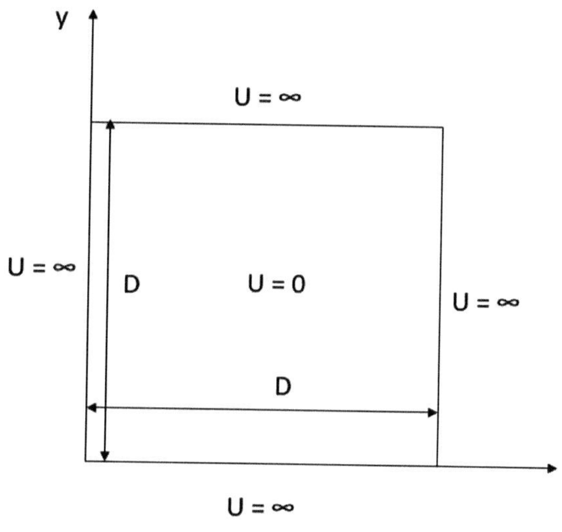

Each term in this equation will be equal to a constant. Therefore,

$$\frac{1}{X}\frac{\partial^2 X}{\partial x^2} = -k_x^2, \tag{2.49}$$

$$\frac{1}{Y}\frac{\partial^2 Y}{\partial y^2} = -k_y^2. \tag{2.50}$$

Here, the constant value is negative because the magnitude of energy should always be positive. Substituting these constants into the Schrodinger wave equation,

$$\frac{2mE}{\hbar^2} = k_x^2 + k_y^2. \tag{2.51}$$

Therefore, solving equations (2.48) and (2.49),

$$X(x) = A\cos k_x x + B\sin k_x x, \tag{2.52}$$

$$Y(y) = C\cos k_y y + D\sin k_y y. \tag{2.53}$$

Substituting the boundary conditions $X(x = 0) = 0$, $X(x = D) = 0$ and $Y(y = 0) = 0$, $Y(y = D) = 0$, we obtain

$$k_x D = n\pi, \tag{2.54}$$

where $n = 1, 2, 3, \ldots, \infty$, and

$$k_y D = l\pi, \tag{2.55}$$

where $l = 1, 2, 3, \ldots, \infty$.

Therefore, substituting k_x, k_y into the energy expression,

$$E_{ln} = \frac{(l^2 + n^2)\pi^2 \hbar^2}{2m D^2}. \tag{2.56}$$

The expression for the wave function can be written as

$$\psi_{ln} = C_{ln} \sin\left(\frac{n\pi x}{D}\right) \sin\left(\frac{l\pi y}{D}\right), \tag{2.57}$$

where $n = 1, 2, 3, \ldots, \infty$ and $l = 1, 2, 3, \ldots, \infty$. Note that this equation has an additional quantization due to the second dimension.

Now, calculate the energy and wave function by considering $n = 1$, $l = 2$, and $n = 2$, $l = 1$. Then, the value of energy remains constant upon flipping the values of n and l. However, the values of the wave function are different. This is because n and l are associated with x and y, respectively. This phenomenon is termed as degeneracy and the energy states associated with these quantum numbers are called degenerate energy levels. In a 1-D confinement, there is no degeneracy because each wave function occupies a given energy level, whereas in a 2-D confinement, more than one wave function exists.

Therefore, in the case of a standing wave, there are no discrete waves and all the waves are continuous, whereas in a 1-D confinement, discrete energy levels exist due to quantization.

2.4.3.1 Pauli's Exclusion Principle

The Schrodinger equation for the 1-D quantum well states that each electron with the given wave function will occupy a particular energy level. However, that is not the complete picture. Pauli's exclusion principle has to be applied to identify the electrons with a spin of $+\frac{1}{2}$ and $-\frac{1}{2}$. According to this principle, no two electrons will have the same set of all four quantum numbers.

2.4.3.2 Vibrational Energy

Attraction or repulsion forces between two atoms can be approximated by a spring–mass system. A repulsive force exists between the two atoms when they are sufficiently close, and an attractive force exists when they are far away. Both attractive and repulsive potentials are shown in Fig. 2.8 by plotting energy on the Y-axis and the distance between the atoms on the X-axis, which is called a potential diagram. Both attractive and repulsive potentials together constitute the Lennard-Jones potential.

However, it is challenging to plug this potential directly into Schrodinger's wave equation. Therefore, an approximation called the harmonic potential model should be used for solving Schrodinger's wave equation. A parabolic potential is assumed

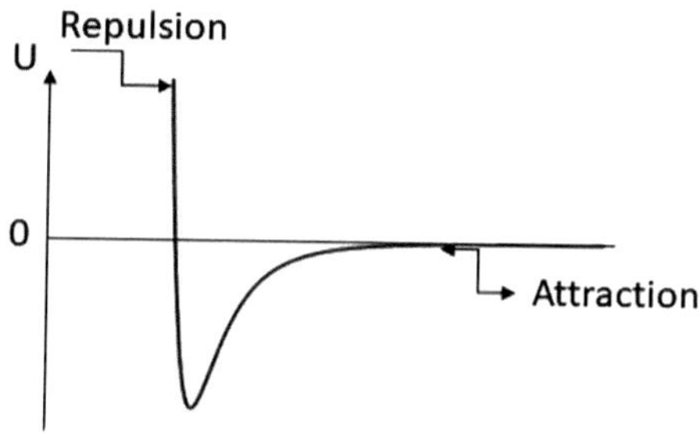

Fig. 2.8 Lennard-Jones potential diagram

to be present at an equilibrium location x_0. If the distance between the two atoms is more than the equilibrium value x_0, an attractive force will be developed, and when the distance between the two atoms is less than x_0, a repulsive force will develop.

After incorporating the potential energy constraint, Schrodinger's wave equation becomes

$$-\frac{\hbar^2}{2m}\frac{\partial^2 \psi}{\partial x^2} + \left(\frac{1}{2}kx^2 - E\right)\psi = 0. \tag{2.58}$$

This is the ODE with non-constant coefficients. The boundary conditions to solve this equation are given by $\psi(x \to -\infty) = 0$ and $\psi(x \to \infty) = 0$.

The solution can be obtained by series expansion, which is given by

$$E_n = h\upsilon\left(n + \frac{1}{2}\right), \tag{2.59}$$

where $n = 0, 1, 2, 3, \ldots, \infty$ and $\upsilon = \frac{1}{2\pi}\sqrt{\frac{k}{m}}$.

The energy at the ground state can be obtained by substituting $n = 0$. Therefore, at $n = 0$, $E_0 = \frac{h\upsilon}{2}$. This means that a certain amount of vibrational energy will exist even if there is no disturbance in the spring–mass system. This energy should have a finite value because it must satisfy Heisenberg's uncertainty principle.

Similarly,

$$\text{at } n = 1, E_1 = \tfrac{3}{2}h\upsilon,$$
$$\text{at } n = 2, E_2 = \tfrac{5}{2}h\upsilon.$$

2.4.3.3 Rotational Energy

There is no rotational energy for a monoatomic system. Therefore, at least a diatomic system has to be taken for analyzing rotational energy. Let us consider two atoms in the XYZ coordinate system, as shown in Fig. 2.9. This system can be considered as a rigid body rotation in the spherical coordinate system in which no relative motion is present between the two atoms. The angle made by these two atoms with the vertical

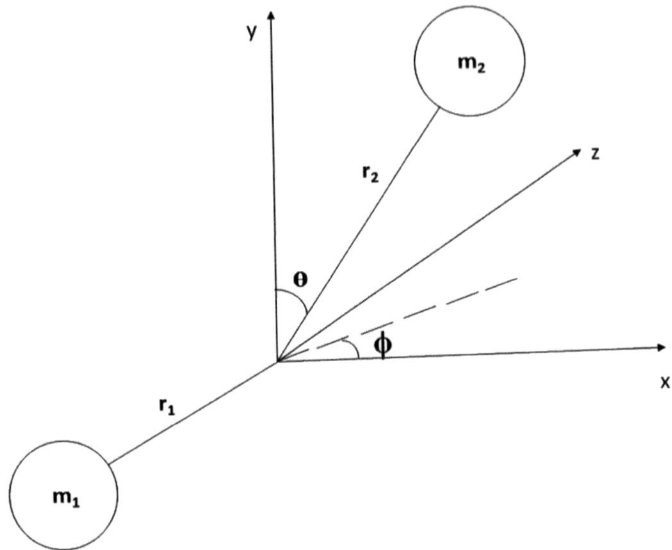

Fig. 2.9 Spherical coordinate system

axis is called the azimuthal angle θ. The projection of this two-atom system in the XZ-plane makes an angle ϕ with the vertical axis. Therefore, ϕ can have values from 0° to 360°.

Let the mass of one atom be m_1 and that of the other be m_2, where $m_1 \neq m_2$. The distance of the first atom from the origin is r_1 and that of the second atom is r_2. The corresponding moment of inertia for this rigid body system is given as

$$I = m_r r_0^2$$

where m_r is called the equivalent mass or reduced mass, which is equal to the harmonic mean of two masses m_1 and m_2:

$$m_r = \frac{m_1 m_2}{m_1 + m_2},$$

and

$$r_0 = r_1 + r_2.$$

On expanding the Schrodinger wave equation in the generic coordinate system,

$$\frac{-\hbar^2}{2m_r} \nabla^2 \psi = E\psi, \tag{2.60}$$

where $U = 0$ because a free electron can just rotate in free space without any potential energy constraints. In the spherical coordinate system,

$$\nabla^2 \psi = \frac{1}{r^2}\left(\frac{\partial}{\partial r}\left(r^2\frac{\partial \psi}{\partial r}\right)\right) + \frac{1}{r^2 \sin\theta}\frac{\partial}{\partial \theta}\left(\sin\theta\frac{\partial \psi}{\partial \theta}\right) + \frac{1}{r^2\sin^2\theta}\frac{\partial^2 \psi}{\partial \phi^2}$$
$$= \frac{-2Emr^2}{\hbar^2}\psi. \tag{2.61}$$

Here, there is no variation concerning r. Hence, the first term can be canceled. The variation in θ and ϕ is considered to describe the rotation of a spherical coordinate system.

By multiplying throughout by r^2 and substituting $mr^2 = I$, the equation becomes

$$\frac{1}{\sin\theta}\frac{\partial}{\partial \theta}\left(\sin\theta\frac{\partial \psi}{\partial \theta}\right) + \frac{1}{\sin^2\theta}\frac{\partial^2 \psi}{\partial \phi^2} = \frac{-2EI}{\hbar^2}\psi, \tag{2.62}$$

where I is the moment of inertia.

The solution of this equation can be obtained by the separation of variables method. Assume that

$$\psi(\theta, \phi) = P(\theta)\psi(\phi). \tag{2.63}$$

Substituting the expression for ψ into the Schrodinger wave equation, we obtain

$$\psi(\phi)\frac{1}{\sin\theta}\frac{\partial}{\partial \theta}\left(\sin\theta\frac{dP}{d\theta}\right) + \frac{1}{\sin^2\theta}P(\theta)\frac{\partial^2 \psi(\phi)}{\partial \phi^2} = \frac{-2EI}{\hbar^2}\psi(\theta, \phi). \tag{2.64}$$

Dividing throughout by ψ, the equation becomes

$$\frac{1}{P(\theta)\sin\theta}\frac{\partial}{\partial \theta}\left(\sin\theta\frac{dP}{d\theta}\right) + \frac{1}{\psi(\phi)\sin^2\theta}\frac{\partial^2 \psi}{\partial \phi^2} = \frac{-2EI}{\hbar^2}. \tag{2.65}$$

This equation can be divided further into two eigenvalue problems by considering the first term as

$$\frac{1}{P(\theta)\sin\theta}\frac{\partial}{\partial \theta}\left(\sin\theta\frac{dP}{d\theta}\right) = -k_\theta^2. \tag{2.66}$$

Moreover, the second term is given as

$$\frac{1}{\psi(\phi)\sin^2\theta}\frac{\partial^2 \psi}{\partial \phi^2} = -m^2. \tag{2.67}$$

Therefore, the equation becomes

$$-k_\theta^2 - m^2 = \frac{-2EI}{\hbar^2}. \tag{2.68}$$

Therefore,

$$k_\theta^2 = \frac{2EI}{\hbar^2} - m^2. \tag{2.69}$$

Substituting k_θ^2 into Eq. 2.65, it becomes

$$\frac{1}{\sin\theta}\frac{\partial}{\partial\theta}\left(\sin\theta\frac{dP}{d\theta}\right) + \left(\frac{2EI}{\hbar^2} - m^2\right)P = 0. \tag{2.70}$$

Solving Eqs. 2.66 and 2.69, the expression for energy (E) becomes

$$E = \frac{\hbar^2}{2I}l(l+1). \tag{2.71}$$

Here, $l \geq |m|$ and $m = 0, \pm1, \pm2, \pm3$ and so on. For example, if $l = 0$, then $m = 0$;

if $l = 1$, then $m = -1, 0, 1.$
if $l = 2$, then $m = -2, -1, 0, 1, 2, 3.$

In general, m can have values from $-l$ to l.

Here, the concept of degeneracy becomes relevant again. If $l = 0$, then $m = 0$. Therefore, two wave functions exist. One is for $l = 0$ and the other is for $m = 0$, but energy is dependent only on l. When $l = 1$, $m = \pm1, 0$. Hence, for $l = 1$, only one value of energy exists, but m can be $\pm1, 0$. Three different wave functions are possible for $m = -1, 0, 1$, respectively. However, all of them have the same value of energy for the rotational system. Therefore, $2l +1$ degenerate energy states exist for each value of l. Each value of l can have a wave function, but the energy value is the same because energy is only a function of l.

Filling of electrons from the lowest energy level to the highest energy level can be predicted with quantum mechanics. Consider an elementary example of the hydrogen atom, as shown in Fig. 2.10. The force of attraction between the nucleus and the electron is called Coulomb's potential.

Coulomb's potential is expressed by

$$F = \frac{1}{4\pi\epsilon_0}\frac{e^2}{r^2}, \tag{2.72}$$

where r is the radius of the hydrogen atom and ϵ is the permittivity of free space.

Coulomb's potential can be written as

$$F = -\frac{dU}{dr}, \tag{2.73}$$

where U is the potential energy. The expression for the potential energy can be obtained by integrating F with respect to r. Therefore,

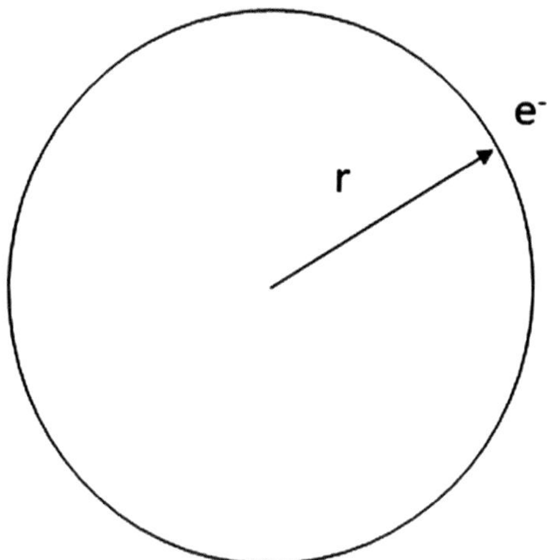

Fig. 2.10 Hydrogen atom

$$U(r) = \frac{-e^2}{4\pi\epsilon_0 r}. \tag{2.74}$$

This can be substituted into Schrodinger's wave equation as a new potential energy constraint. The expression for electronic energy obtained by solving Schrodinger's wave equation by the separation of variables method is given by

$$E_n^{el} = \frac{-mC_1^2}{2\hbar^2 n^2} = \frac{-13.6\,ev}{n^2}, \tag{2.75}$$

where m is the mass of the electron, C_1 is a constant, and n is the principal quantum number. Here, l, m, and n are the quantum numbers corresponding to the r, θ, and ϕ directions, respectively. n should be greater than or equal to 1 ($n \geq 1$), l can have values from 0 to $n - 1$ and m can have values from $-l$ to l.

The concept of degeneracy becomes relevant again. The values of l, m, ψ for $n = 1, 2$ are shown in Table 2.1. From the table, for $n = 2$, we see that four different wave functions exist with the same value of energy. This is the classical example of degeneracy.

The electronic configuration can be determined based on the quantum numbers. The quantum numbers $l = 0, 1, 2, 3$, and 4 correspond to s, p, d, f and g orbitals, respectively. For example, the combination of $n = 1, l = 0$, and $m = 0$ corresponds

Table 2.1 The values of l, m, and ψ corresponding to $n = 1, 2$

n	l	m	ψ	E (W)
1	0	0	ψ_{100}	-13.6
2	0,1	$-1, 0, 1$	$\psi_{200}, \psi_{201}, \psi_{210}, \psi_{211}$	-3.4

to $1s$ orbital. Here, $n = 2$ corresponds to $2s$, $2p$ orbitals. Six quantum states corre-spond to $2p$ and two quantum states correspond to $2s$ or, in other words, six electrons can occupy $2p$ orbital and two electrons can occupy $2s$ orbital. The electronic config-uration can be written as $2s^2 2p^6$. Similarly, the electronic configuration for $n = 3$ is $3s^2 3p^6 3d^{10}$. When an electron jumps from a higher energy state to a lower quantum state, emission of energy takes place, which is given by

$$E = h\upsilon. \tag{2.76}$$

Therefore,

$$h\upsilon = 13.6 \left(\frac{1}{n_2^2} - \frac{1}{n_1^2} \right) ev. \tag{2.77}$$

2.4.4 *1-D Quantum Well with a Finite Potential Barrier Height V*

Consider a 1-D quantum well having a finite potential barrier height V, as shown in Fig. 2.11. The width of the well is D. There is no potential energy constraint inside the well, which means that the particle can freely move inside the well.

For $x < 0$ or $x > D$, the potential energy constraint is finite and is equal to V. In the case of an infinite potential well, the wave function outside the well is 0. Therefore, only one solution exists. However, in this case, the electron can come out

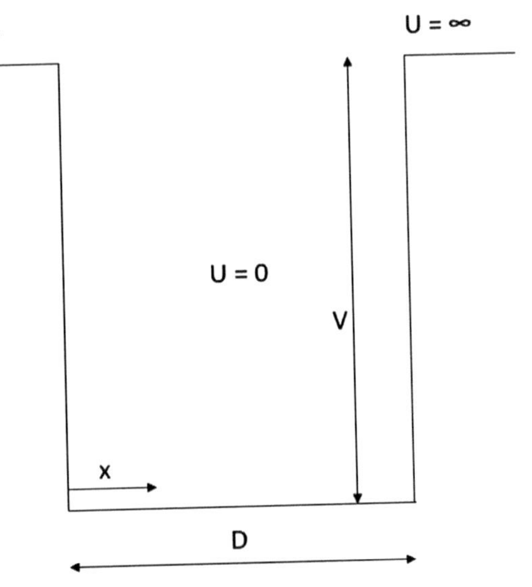

Fig. 2.11 1-D quantum well having a finite potential barrier height V

of the well. Hence, the solution exists at $x < 0$ or $x > D$. Therefore, totally three solutions exist.

Writing Schrodinger's wave equation for $0 < x < D$,

$$\frac{-\hbar^2}{2m}\frac{d^2\psi}{dx^2} - E\psi = 0. \tag{2.78}$$

The solution of this equation can be written as

$$\psi(x) = Be^{-kx} + Ce^{kx}. \tag{2.79}$$

The solution is a combination of both right-running and left-running waves, that is, electrons run toward the right or left constraint inside the potential well. Writing Schrodinger's wave equation at $x < 0$,

$$\frac{-\hbar^2}{2m}\frac{d^2\psi}{dx^2} - (E - V)\psi = 0. \tag{2.80}$$

The solution can be written as

$$\psi_1(x) = Ae^{k_1 x} + A^1 e^{-k_1 x}, \tag{2.81}$$

where $\psi = 0$ when x tends to $-\infty$. Therefore $A^1 = 0$ and the solution becomes

$$\psi_1(x) = Ae^{k_1 x}, \tag{2.82}$$

where $k_1 = \sqrt{\frac{2m(V-E)}{\hbar^2}}$. Similarly, for $x > D$, the solution can be written as

$$\psi_3(x) = D^1 e^{k_3 x} + De^{-k_3 x}. \tag{2.83}$$

On applying the condition $x \to \infty$, $D^1 = 0$. Now, the solution becomes

$$\psi_3(x) = De^{-k_3 x}, \tag{2.84}$$

where $k_3 = \sqrt{\frac{2m(V-E)}{\hbar^2}}$.

Note that at $x = 0$, $\psi_1 = \psi_2$. Similarly, at $x = D$, $\psi_2 = \psi_3$. Therefore, on applying $\psi_1 = \psi_2$ at $x = 0$, we obtain $A = B + C$. The slopes are also equal at $x = 0$, which gives

$$\frac{\partial \psi_1}{\partial x} = \frac{\partial \psi_2}{\partial x} \Rightarrow k_1 A = ik_2(B - C). \tag{2.85}$$

Similarly, on applying $\psi_2 = \psi_3$ at $x = D$, we obtain

$$\psi_2 = \psi_3 \Rightarrow Be^{ik_2 D} + Ce^{-ik_2 D} = De^{-k_1 D}. \tag{2.86}$$

Note that

$$k_1 = k_3.$$

Similarly,

$$\frac{\partial \psi_2}{\partial x} = \frac{\partial \psi_3}{\partial x}. \tag{2.87}$$

On solving these equations, the required values of energy can be obtained.

A brief overview of quantum mechanics is given in this chapter to understand the wave effects at the nano level. The Schrodinger's wave equation, which is the most fundamental equation for describing the quantum mechanical behavior of waves, was solved and the concepts of a free particle in a 1-D, 2-D confinement, 1-D quantum well with a finite potential barrier height V are illustrated clearly.

Exercise Problems

1. For the $n = 2$ state of an electron inside an infinite potential well, prove that the Heisenberg uncertainty relation is satisfied.
2. Plot the most probable electron distribution in a one-dimensional infinite potential well for $n = 1, 2, 3$.
3. The fundamental vibrational frequency of the H_2 molecule is 4401 cm^{-1} and its rotational constant is 59.32 cm^{-1}. Determine the photon emission wavelengths due to the combined vibrational-rotational modes in H_2 near the fundamental vibrational mode.
4. The minimum focal point of the electron beam used in a Transmission electron microscope (TEM) depends on its wavelength. Determine the electron wavelength if they have an energy of (a) 100 keV and (b) 1 MeV.
5. Derive expressions to determine the electron energy levels in a potential well surrounded by a barrier of finite height, as shown in the figure below. For $D = 50$ Å, determine the first three energy levels for $V = 0.5$ eV and 1 eV.
6. Consider an electron of energy E moving from left to right as shown in the figure below. It encounters a potential barrier of height V. The electron wave can either be reflected or transmitted

    (a) Write the expressions for the incoming, reflected and transmitted wave functions.
    (b) At the interface $x = 0$, the wave function and its first derivative must be continuous. Using these conditions, derive expressions for the ratio of constants B/A and C/A appearing in the expressions for wave functions of part (a).
    (c) The reflectivity R is defined as the ratio of the reflected particle flux divided by the incoming particle flux as $R = Jr/Ji$, and similarly for transmissivity as $T = Jt/Ji$. Derive expressions for R and T.

(d) For $E = 1.2$ eV and $V = 1$ eV, calculate the electron reflectivity and transmissivity.

(e) For $E = 0.8$ eV and $V = 1$ eV, show that the transmissivity is zero, and also show that the wave function is non zero. This nonzero wave function that does not carry any flux is called an evanescent wave.

Chapter 3
Fundamentals of Solid-State Physics

3.1 Crystal Structure and Definitions

Solid-state physics describes the actual behavior of electrons in a crystal structure. In this field, the free electron model is extended to a solid crystal structure. The concept of lattice becomes relevant here.

Lattice
A lattice is the regular arrangement of atoms in space separated by a fixed distance in metals or a crystalline solid. Unit cells can be constructed based on lattices.

Unit Cell
A unit cell is the smallest group of atoms with the overall symmetry of a crystal structure. Continuous repetition of unit cells in 3-D space results in the formation of a crystal structure, as shown in Fig. 3.1. Therefore, the periodicity principle can be applied to a unit cell.

The electric potential between the atoms inside a crystal structure is periodic. Therefore, a unit cell is used to model the motion of an electron. The electron is constrained by the potential due to the interaction between the atoms. While quantum mechanics focuses on the analysis of a single electron in a quantum well, solid-state physics deals with the analysis of the free electrons subjected to an arbitrary potential inside the crystal by using the Kronig–Penney model.

3.2 Kronig–Penney Model

The Kronig–Penney model simplifies more complex interaction potentials into simple rectangular periodic potentials. The width of a quantum well is a, and it is separated by small potential barriers of width b, as shown in Fig. 3.2. The solution of the Schrodinger wave equation for the periodic potential is repetitive in the entire crystal.

© The Author(s) 2025
A. Pattamatta and S. K. Das, *Fundamentals of Nano- and Microscale Heat Transport*,
https://doi.org/10.1007/978-3-031-89613-2_3

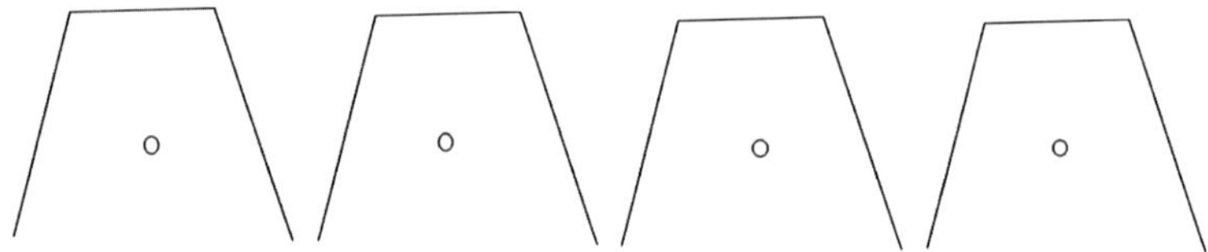

Fig. 3.1 Crystal structure

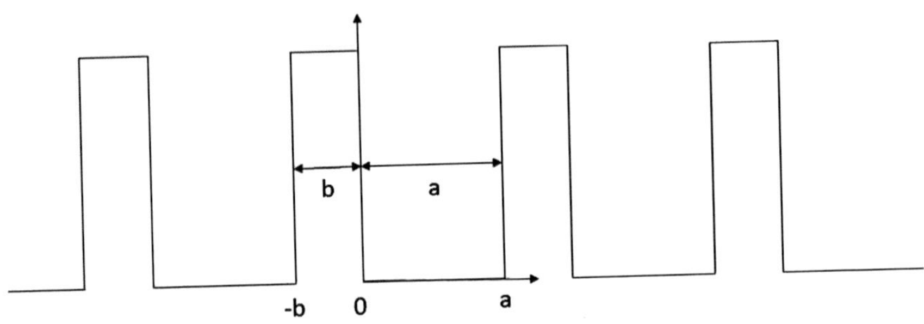

Fig. 3.2 Kronig–Penney model

The periodicity of this potential is $a + b$. A coordinate system can be established at one end of the potential barrier. Hence, the coordinate system from $-b$ to 0 is substantially the potential barrier width and the coordinate system from 0 to a is the quantum well width. This is the representation of a single unit cell.

Therefore, the analysis of a single periodic unit cell potential can be carried out, and then, the periodic boundary condition can be applied. The boundary condition at $x = 0$ should also be given because this is where the transition occurs from the confined electron within the potential well to the free electron.

The potential distribution is given as follows:

$$U(x) = \begin{cases} 0 & \text{for } 0 < x < a \\ U_0 & \text{for } -b < x < 0. \end{cases}$$

Therefore, the Schrodinger wave equation for the periodic potential using these constraints can be written as follows:

$$-\frac{\hbar^2}{2m}\frac{\partial^2 \psi}{\partial x^2} - E\psi = 0, \quad 0 < x < a, \tag{3.1}$$

$$-\frac{\hbar^2}{2m}\frac{\partial^2 \psi}{\partial x^2} + (U_0 - E)\psi = 0, \quad -b < x < 0. \tag{3.2}$$

The solution of the Schrodinger equation inside the quantum well ($0 < x < a$) is given by

$$\psi_1(x) = Ae^{ikx} + Be^{-ikx}. \tag{3.3}$$

The expression for energy is given by

$$E = \frac{-\hbar^2 k^2}{2m}. \tag{3.4}$$

The solution inside the potential barrier $(-b < x < 0)$ is given by

$$\psi_2(x) = Ce^{Qx} + De^{-Qx}. \tag{3.5}$$

The energy expression can be obtained from the following equation:

$$(U_0 - E) = \frac{\hbar^2 Q^2}{2m}. \tag{3.6}$$

At $x = 0$, $\psi_1(x) = \psi_2(x)$ and $\psi_1^1(x) = \psi_2^1(x)$. The wave function corresponding to the periodic potential can be obtained using the Bloch theorem.

The Bloch theorem states that

$$\psi(x + (a + b)) = \psi(x)e^{ik(a+b)}, \tag{3.7}$$

where

$$\kappa = \frac{2\pi n}{l} \quad (n = 0, \pm 1, \pm 2, \pm 3, \dots), \tag{3.8}$$

where l is the length of the crystal $(l = N(a + b))$ and n is the number of atoms inside the crystal.

Therefore, on applying the Bloch theorem at $x = a$,

$$\psi(x = a) = \psi(x = -b)e^{ik(a+b)}. \tag{3.9}$$

This indicates that the wave function at $x = a$ can be written in terms of $x = -b$ because the potential is periodic about $a + b$.

On substituting the wave functions at $x = a$, $x = -b$ into the expression, we obtain

$$Ae^{i\kappa a} + Be^{-i\kappa a} = (Ce^{-Qb} + De^{Qb})e^{i\kappa(a+b)}. \tag{3.10}$$

On applying $\psi^1(x = a)$, we obtain

$$\psi^1(x = a) = \psi^1(x = -b)e^{i\kappa(a+b)}$$
$$\Rightarrow i\kappa(Ae^{ika} - Be^{-i\kappa a}) = Q(-Ce^{-Qb} + De^{Qb})e^{i\kappa(a+b)}. \tag{3.11}$$

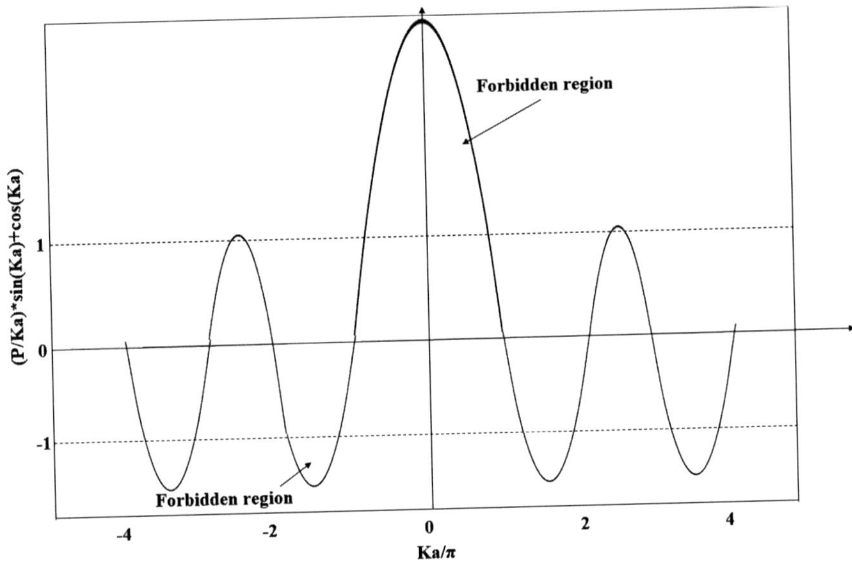

Fig. 3.3 $\frac{ka}{\pi}$ versus $\frac{P}{ka}\sin ka + \cos ka$

From these equations, one can write

$$\frac{P}{\kappa a} + \sin \kappa a + \cos \kappa a = \cos \kappa a, \tag{3.12}$$

where

$$P = \frac{Q^2 ba}{2}. \tag{3.13}$$

Therefore, the solution of this equation can be obtained by plotting $\frac{ka}{\pi}$ on the x-axis and $\frac{P}{ka}\sin ka + \cos ka$ on the y-axis, as shown in Fig. 3.3. Here, $\frac{ka}{\pi}$ is a non-dimensional wave number which can take negative values. On plotting, one can observe that the curve is symmetric about 0. Also, it has a peak value at 0.

The term $\cos \kappa a$ term on the right-hand side has maximum and minimum values at $+1$ and -1, respectively. The region corresponding to $\cos \kappa a$ should be identified in the graph. The solution exists in the common region corresponding to the terms on both the right-hand and left-hand sides. The regions above $+1$ and below -1 are called the forbidden regions. These values cannot satisfy the equation because they do not belong to the common region. Therefore, there exist no electrons in that region.

To calculate the energy of the electrons, the κ value should be determined by plotting the vertical lines in the graph. The value of energy is 0 in the forbidden region since no electrons exist in this region. These forbidden regions produce band gaps, on the basis of which metals, semiconductors, and insulators can be classified.

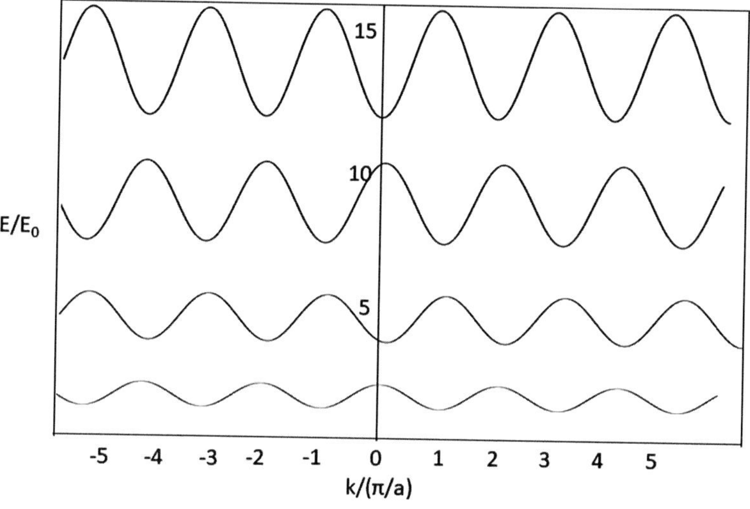

Fig. 3.4 Dispersion curve

The solutions of κ are discrete values but not continuous because $\kappa = \frac{2\pi n}{L}$, which results in the band gap or separation between the energy bands. Therefore, this plot can be converted into a plot of energy *vs.* κ and is called the dispersion curve. Here, the quantity $\frac{ka}{\pi}$ is plotted on the x-axis and the non-dimensional energy ($\frac{E}{E_0}$) is plotted on the y-axis, as shown in Fig. 3.4. Note that the energy (E) and the crystal wave number κ can be non-dimensionalized by dividing by E_0 and π/a, respectively, where

$$E = \frac{\hbar^2}{8ma^2}.$$

The dispersion of energy in the entire crystal can be analyzed by plotting the dispersion curve. The energy values corresponding to various levels can be plotted in a dispersion curve. The dispersion curve shows that the energy gap or the band gap exists between successive energy states.

Note that all these energy states are periodic because the periodic potential exists inside the crystal structure.

3.3 Classification of Solids: Insulators, Conductors, and Semiconductors

Consider a unit cell between the limits -1 and $+1$. From the graph, we can see that the dispersion curve is periodic in this region. Also, it is symmetric about 0 between the limits -1 and $+1$, as shown in Fig. 3.5. Consider one half of a unit cell

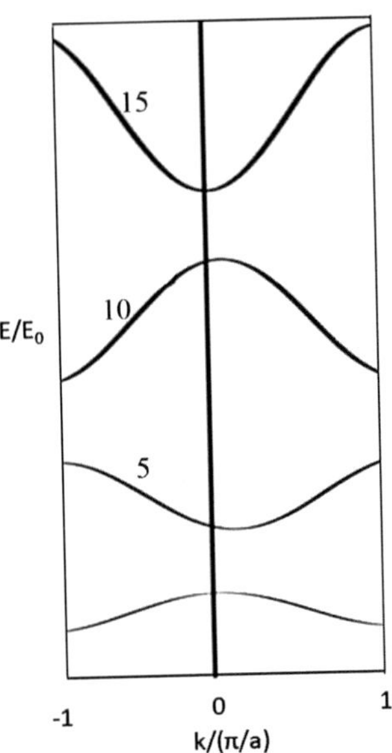

Fig. 3.5 Dispersion curve between the limits -1 and $+1$

for analysis, which is merely a mirror image of the other side. For example, consider the positive quadrant such that $\frac{ka}{\pi}$ varies from 0 to 1 and is plotted on the x-axis and $\frac{E}{E_0}$ is plotted on the y-axis.

In metals, the electrons first occupy all the lower energy levels. Then they will start occupying the next such higher energy levels. At $0\ K$, the electrons attain the maximum energy level called the Fermi level. The Fermi energy level is very close to the conduction band. As temperature increases, these electrons can freely move within the conduction band. In the conduction band, there is plenty of space available for the free electrons to move. Due to this property, metals are considered to be excellent conductors of heat and current.

In insulators, the electrons first occupy the lowest energy levels and then they occupy the Fermi energy level at 0 K. But, the band gap is more. This indicates that the valence band is filled with electrons and the conduction band is empty. Therefore, no free electrons are present in the conduction band.

Figure 3.6 represents the dispersion curve for both metals and insulators. In general, a material having a band gap greater than 3 ev is an insulator and a material having a band gap less than 3 ev is a semiconductor.

Semiconductors can be classified into intrinsic and extrinsic types. Figure 3.7 shows the dispersion curve for intrinsic, n-type, and p-type semiconductors.

A p-type or an n-type semiconductor can be classified based on a donor or an acceptor whose Fermi levels can be tuned to be close to either the valence band or the conduction band.

In an n-type semiconductor, the Fermi energy of the donor is closer to the conduction band of the material. Therefore, at a higher temperature, the electrons from

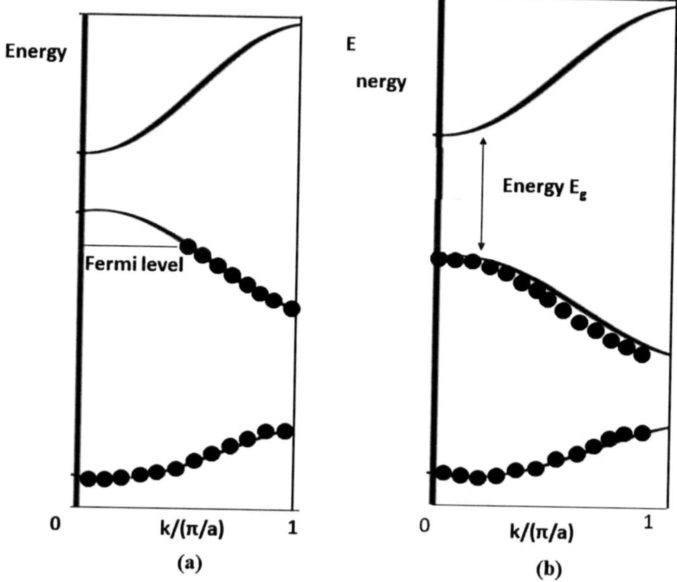

Fig. 3.6 **a** Metals; **b** insulators

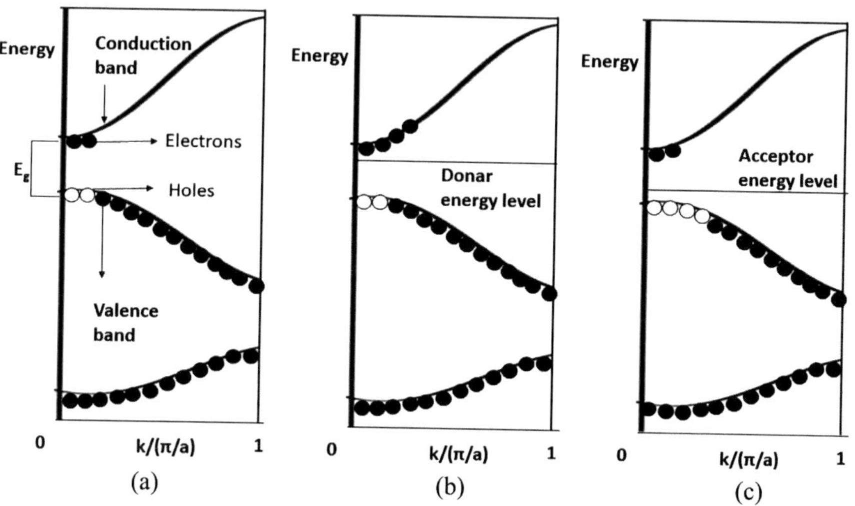

Fig. 3.7 **a** Intrinsic semiconductor; **b** *n*-type semiconductor; **c** *p*-type semiconductor

the donor jump to the conduction band of the acceptor material and this generates an *n*-type semiconductor. Phosphorous is an example of an *n*-type semiconductor.

In a *p*-type semiconductor, the Fermi level of an acceptor is closer to the valence band of the material.

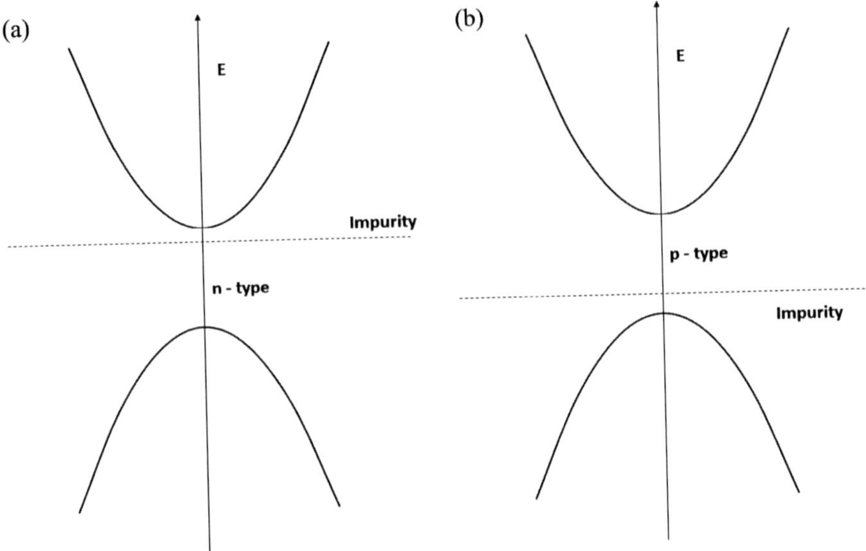

Fig. 3.8 **a** n-type semiconductor; **b** p-type semiconductor

The valence band and the conduction band can be represented graphically with two parabolas symmetric about the x-axis, as shown in Fig. 3.8. The space between these two parabolas is the band gap. In an n-type semiconductor, the donor will have a Fermi level closer to the conduction band, whereas in a p-type semiconductor, the acceptor will have a Fermi level closer to the valence band. The movement of the electrons is close to either the maxima of the lower band or the minima of the upper band.

Therefore, the energy at an arbitrary wave vector can be calculated through Taylor's series approximation. Here, k_m is the location where the maxima or minima for the wave vector exists. Expanding the energy about k_m by Taylor's series,

$$E(k) = E(k_m) + \frac{\partial E}{\partial k}\bigg|_{k_m} (k - k_m) + \frac{1}{2} \frac{\partial^2 E}{\partial k^2}\bigg|_{k_m} (k - k_m)^2 + \cdots. \tag{3.14}$$

Therefore, at maxima or minima,

$$\frac{dE}{dk}\bigg|_{k_m} = 0.$$

Now, the equation becomes

$$E(k) = E(k_m) + \frac{1}{2} \frac{\partial^2 E}{\partial k^2}\bigg|_{k_m} (k - k_m)^2. \tag{3.15}$$

Here, the effective mass or crystal mass is given by

$$m^* = \frac{\hbar^2}{\left.\frac{\partial^2 E}{\partial k^2}\right|_{k_m}} . \tag{3.16}$$

m^* establishes the relationship between the energy and the crystal wave vector k. Substituting the crystal mass m^* into the Taylor series expansion,

$$E(k) = E(k_m) + \frac{1}{2}\frac{\hbar^2}{m^*}(k - k_m)^2. \tag{3.17}$$

This is the dispersion relation for the entire crystal.

Applying this dispersion relation to metals gives

$$E = E_c + \frac{1}{2}\frac{\hbar^2 k^2}{m^*}. \tag{3.18}$$

Here, E_c is the energy corresponding to the conduction band where minimum energy exists.

In the case of a 1-D quantum well, as shown in Fig. 3.9, all the energy levels are quantized.

The expression for the wave vector k is given by

$$k = \frac{n\pi}{D}, \tag{3.19}$$

Fig. 3.9 1-D quantum well

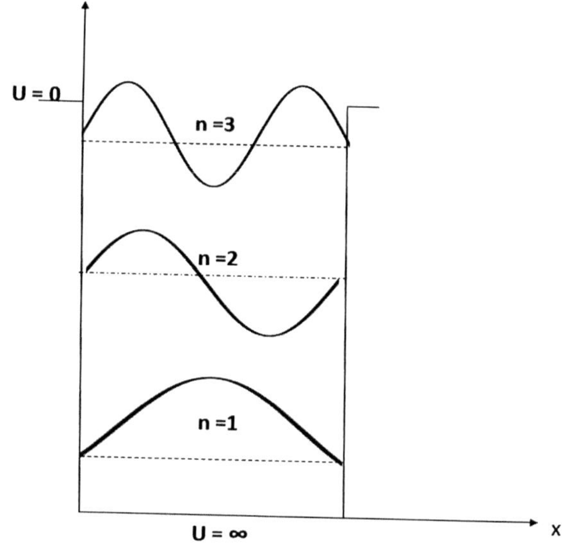

where D is the width of the well. Discrete energy values can be observed corresponding to $n = 1, 2, 3, \ldots$ In a crystal structure, two levels of quantization can be observed. One is due to the wave vector $k = \frac{2\pi n}{L}$, where $n = \pm 1, \pm 2, \pm 3, \ldots$, and the other is due to the solution of the Schrodinger equation, which states that there are non-allowable solutions called band gaps.

3.4 Phonon Dispersion

In quantum mechanics, a single-particle vibrational energy state is analyzed by using the spring–mass system. In solid-state physics, vibrational waeves are analyzed collectively for a large number of atoms. These vibrational waves are termed phonons. Phonons can be analyzed by Newtonian mechanics.

3.4.1 1-D Monoatomic Lattice Chain

A simple model can be constructed for analysis, as shown in Fig. 3.10. The interaction between the atoms in a crystal is assumed to be connected by springs with a spring constant k. The separation between the atoms can be denoted by a. As this is a 1-D representation of the atoms in a crystal, the total length of a crystal having N atoms is given by $(N - 1) \cdot a$ or simply Na. There is no difference between using N or $N - 1$ since the total number of atoms in a crystal is of the order of 10^{26} or 10^{28}.

The location of an arbitrary atom inside the crystal can be denoted by ja. Therefore, the location of an adjacent right atom is $(j + 1)a$ and that of an adjacent left atom is $(j - 1)a$.

Consider an atom at a location ja. Two forces F_1 and F_2 are acting on this atom in opposite directions, as shown in Fig. 3.11. The effective or resultant force is equal to the algebraic sum of the forces acting on the atom. The force acting on the atom in the right direction F_1 is $k(u_{j+1} - u_j)$ and that acting on the atom in the left direction F_2 is $k(u_j - u_{j-1})$, where u is the displacement and k is the spring constant. Therefore,

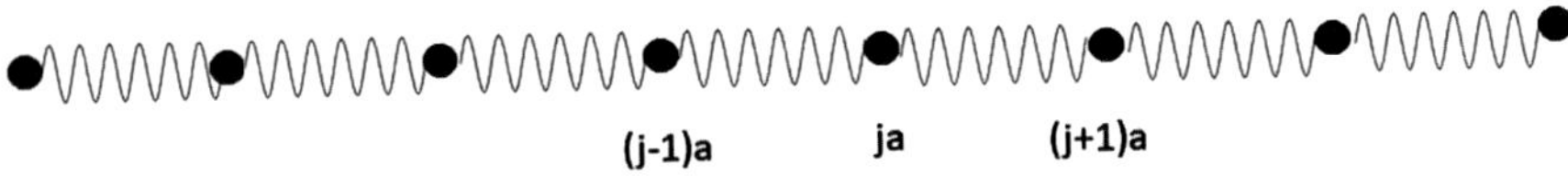

$(j-1)a$ ja $(j+1)a$

Fig. 3.10 1-D monoatomic lattice chain

Fig. 3.11 An arbitrary atom inside the monoatomic lattice chain

the net force acting on the atom is $F = k(u_{j+1} - 2u_j - u_{j-1})$. In terms of a wave picture, it produces a right-running wave.

The resultant force can be written as

$$m \frac{d^2 u_j}{dt^2} = k(u_{j+1} - 2u_j - u_{j-1}). \tag{3.20}$$

Multiplying and dividing by a^2 on the right-hand side, we obtain

$$m \frac{d^2 u_j}{dt^2} = \frac{ka^2(u_{j+1} - 2u_j - u_{j-1})}{a^2}. \tag{3.21}$$

This equation can be written as

$$m \frac{d^2 u_j}{dt^2} = ka^2 \frac{\partial^2 u_j}{\partial x^2}, \tag{3.22}$$

which is the hyperbolic wave equation.

The solution of this wave equation is a right-running wave, which can be expressed as

$$u_j = Ae^{-i(wt-kx)}. \tag{3.23}$$

However, this is a continuous solution. Therefore, this can be converted into a discrete solution by substituting $x = ja$ into the above equation, i.e.,

$$u_j = Ae^{-i(wt-kja)}. \tag{3.24}$$

Differentiating twice with respect to x and t, we obtain

$$\frac{\partial u}{\partial t} = Ae^{ikja}.e^{-i\omega t}(-i\omega), \tag{3.25}$$

$$\frac{\partial^2 u}{\partial t^2} = Ae^{ikja}.e^{-i\omega t}(-i\omega)(-i\omega). \tag{3.26}$$

Substituting into Eq. (3.22), we obtain

$$m[Ae^{ikja}(-i\omega)^2 e^{-i\omega t}] = k[Ae^{-i(\omega t - k(j+1)a)}$$
$$- 2Ae^{-i(\omega t - kja)} + Ae^{-i(\omega t - k(j-1)a)}]. \tag{3.27}$$

Further simplification yields the following:

$$\begin{aligned} -\omega^2 m &= k(e^{ika} + e^{-ika} - 2) \\ &= 2k(\cos ka - 1) \\ &= -2k\sin^2 \frac{ka}{2}. \end{aligned} \tag{3.28}$$

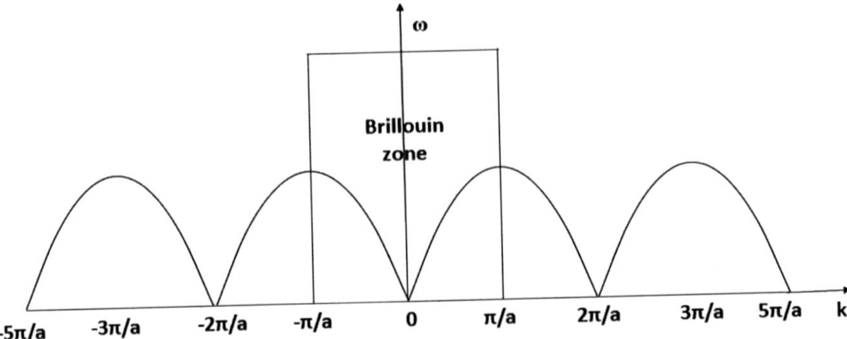

Fig. 3.12 Graphical representation of the dispersion relation

Therefore,

$$\omega = 2\sqrt{\frac{k}{m}}\left|\sin\left(\frac{ka}{2}\right)\right|, \tag{3.29}$$

where $k = \frac{2\pi n}{L}$. This is called the dispersion relation. Here, two levels of discreteness exist due to n and m, where $n = 0, 1, 2, \ldots$ and $m = \pm 1, \pm 2, \pm 3, \ldots$.

The energy can be expressed as

$$E_n = \hbar\omega\left(n + \frac{1}{2}\right), \tag{3.30}$$

where $n = 0, 1, 2, \ldots$

The dispersion relation can be represented graphically by plotting k on the x-axis and ω on the y-axis, as shown in Fig. 3.12. ω will have the maximum value at $k = \frac{\pi}{a}, \frac{3\pi}{a}, \ldots$. The dispersion curve is periodic with a period $\frac{2\pi}{a}$. The block between $k = -\frac{\pi}{a}$ and $\frac{\pi}{a}$ is called the Brillouin zone. This zone is the unit cell of the crystal about which the dispersion or energy relationship is periodic.

The vibration of the phonon waves can be analyzed by considering the 1-D monoatomic chain of atoms in a crystal, as shown in Fig. 3.13.

These atoms are separated by the lattice constant a. Therefore, the minimum wavelength for the phonon vibrational waves is equal to $2a$, as shown in the following equation:

$$\lambda_{\min} = 2a. \tag{3.31}$$

The minimum vibrational wavelength cannot be smaller than $2a$ because there are no atoms. Therefore,

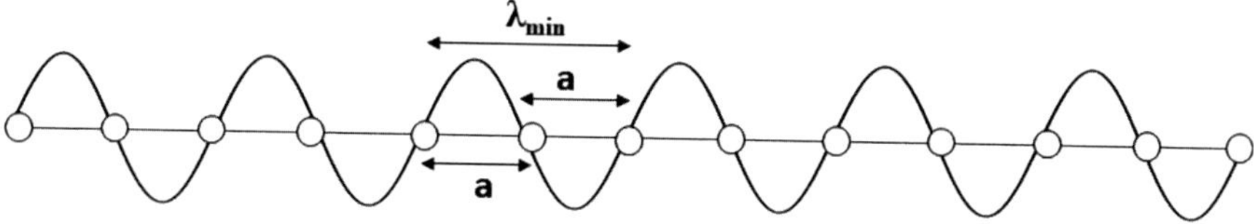

Fig. 3.13 1-D monoatomic chain of atoms in a crystal

$$\lambda > 2a \Rightarrow \frac{\lambda}{2\pi} > \frac{a}{\pi}$$

$$\Rightarrow \frac{2\pi}{\lambda} \leq \frac{\pi}{a} \tag{3.32}$$

$$\Rightarrow k \leq \frac{\pi}{a}.$$

Hence, the entire solution is confined to the Brillouin zone because $k \leq \frac{\pi}{a}$. We have

$$k = \frac{2\pi m}{Na}, \tag{3.33}$$

where $m = \pm 1, \pm 2, \pm 3, \dots$

The relationship between m and N can be established by equating Eqs. (3.32) and (3.33). Therefore,

$$Fm \leq \frac{N}{2}. \tag{3.34}$$

The number of allowable quantum states from $k = 0$ to $\frac{\pi}{a}$ is less than or equal to half of the number of atoms in a crystal. Therefore, within the Brillouin zone, i.e., from $-\frac{\pi}{a}$ to $+\frac{\pi}{a}$, the maximum allowable number of quantum states is equal to N. In general, the waves are longitudinal. The slope of the dispersion curve, $\frac{d\omega}{dk}$, gives the wave velocity. Wave velocity is defined as the speed of sound in the lattice since longitudinal waves correspond to acoustic waves. Therefore, to calculate the acoustic velocity, the slope of the dispersion curve, which gives the speed of the longitudinal phonons, should be known.

Now, the analysis of longitudinal vibrational modes can be extended to 3-D systems. In 3-D systems, longitudinal waves propagate along the x-direction and transverse waves propagate along the y- and z-directions, respectively.

Therefore, in 3-D systems, the dispersion curves for both longitudinal and transverse waves can be obtained by plotting ω vs. k, as shown in Fig. 3.14. The transverse waves will also contribute to dispersion. Therefore, they will have a separate set of dispersion curves. The transverse waves will also have the slope corresponding to certain velocities. The velocities of these transverse waves are not equal to the speed of sound but correspond to the speed of transverse phonons. These transverse vibrations also contribute to the heat transport phenomenon along with longitudinal vibrations.

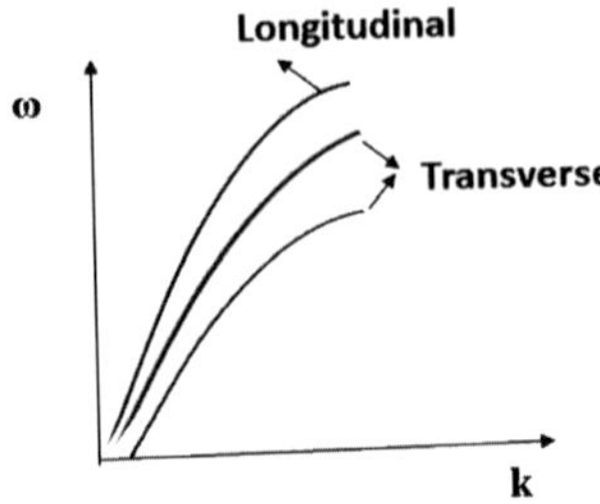

Fig. 3.14 Dispersion curves for both longitudinal and transverse waves

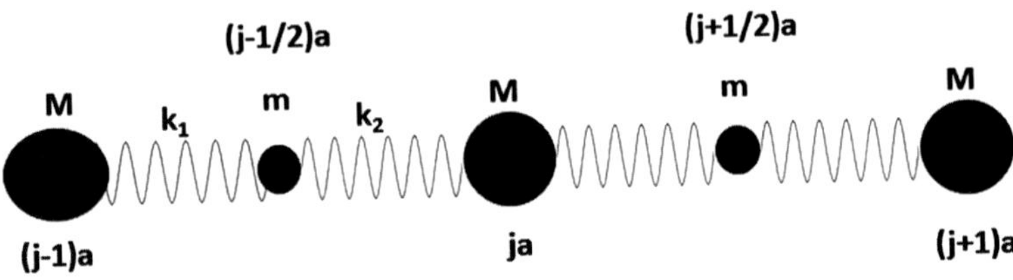

Fig. 3.15 Polyatomic lattice chain

Therefore, the dispersion curve is important to not only understand how frequency varies with wave vector but also analyze heat transport phenomenon.

3.4.2 1-D Polyatomic Lattice Chain

So far, the monatomic atoms in a crystal have been analyzed in detail. Henceforth, this analysis will be extended to a polyatomic lattice chain. Consider a polyatomic chain having an alternate arrangement of both larger and smaller atoms, as shown in Fig. 3.15. The masses of the larger and smaller atoms are M and m, respectively. The alternate arrangement of larger and smaller atoms together forms a polyatomic crystal structure. The lattice constant between the larger and smaller masses is K_1 and that between the smaller and larger masses is K_2. Therefore, two different lattice constants K_1 and K_2 repeat alternately throughout the entire crystal. The spacing between two successive larger masses is a. This spacing "a" can be divided into the spacing between the larger mass, smaller mass a_1 and the smaller mass, larger mass a_2. Therefore, $a_1 + a_2 = a$ and $a_1 = a_2$. The larger mass is at location ja and the adjacent larger masses are at locations $j + 1a$, $j - 1a$, respectively. The smaller masses adjacent to the larger mass are at locations $j + \frac{1}{2}a$, $j - \frac{1}{2}a$, respectively. The spring constants k_1 and k_2 are equal to each other. Therefore, $k_1 = k_2 = k$. The relative displacements from equilibrium for the larger and smaller masses are at u and v, respectively.

The force balance equations for both larger (M) and smaller (m) masses are written as follows:

$$M_i = M\frac{d^2 u_j}{dt^2} = k(v_{j+\frac{1}{2}a} - 2u_j + v_{j-\frac{1}{2}a}), \tag{3.35}$$

$$m_i = m\frac{d^2 v_{j+\frac{1}{2}a}}{dt^2} = k(u_{j+1a} - 2v_{j+\frac{1}{2}a} + u_{ja}). \tag{3.36}$$

The solutions for u_j, $v_{j+\frac{1}{2}}$ are assumed to be

$$u_j = Ae^{-i(\omega t - Kja)}, \tag{3.37}$$

$$v_{j+\frac{1}{2}} = Be^{-i(\omega t - K(j+\frac{1}{2}a))}. \tag{3.38}$$

These solutions are right-running waves. Therefore, the final solution is given by

$$\omega^2 = K\left(\frac{1}{M} + \frac{1}{m}\right) \pm K\left[\left(\frac{1}{M} + \frac{1}{m}\right)^2 - 4\frac{\sin^2(Ka/2)}{Mm}\right]^{1/2}. \tag{3.39}$$

This is the dispersion solution for the polyatomic chain of atoms. Therefore, effectively the waves from the larger and smaller atom displacements exist inside the crystal structure, which can be either out of phase or in phase with each other. The positive sign denotes the out-of-phase component and the negative sign denotes the in-phase component. The out-of-phase component shown in Fig. 3.16 consumes more energy than the in-phase component because both waves corresponding to larger and smaller masses do not overlap.

The out-of-phase wave is also called the optical mode or optical branch of phonons and the in-phase wave is called the acoustic branch. The slope of the acoustic wave provides the speed of sound in the lattice structure. The out-of-phase component corresponds to the higher energy mode of phonon vibration, which is usually associated with the interaction of photons. The phonon–photon coupling can be explained with the example of a semiconductor solar cell exposed to UV radiation. Absorption of photons takes place after it is exposed to UV radiation. Some heat is also generated within the semiconductor material due to the photon–phonon interactions. In this process, the optical branch interacts with the photon light waves and transfers the energy to the acoustic branch, which can then be dissipated in the form of heat. This is a very complex phenomenon and is not yet very clearly understood.

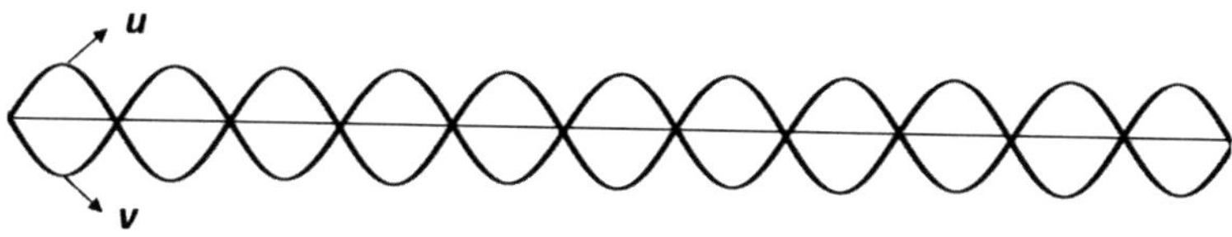

Fig. 3.16 Out-of-phase component

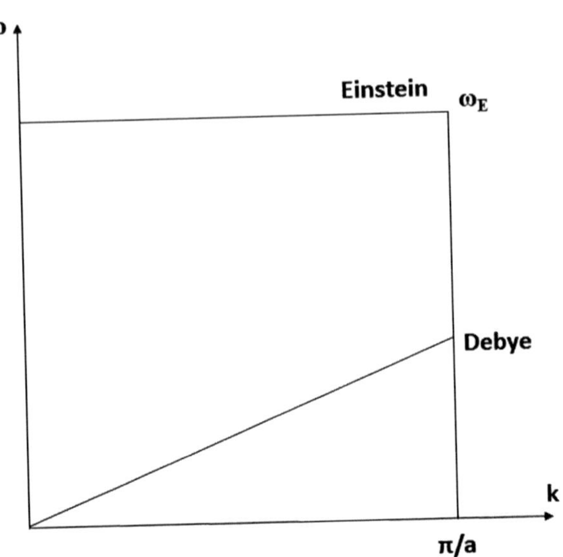

Fig. 3.17 Representation of Debye and Einstein approximations

The total number of acoustic branches in a 3-D crystal structure is 3, irrespective of the number of atoms. However, the number of optical branches is $3(m - 1)$, where m is the number of atoms in a crystal. Experimental measurements can help examine the validity of dispersion curves. Researchers have found that experimental measurements are in good approximation with the values of dispersion curves. However, to further simplify the dispersion analysis, Debye approximation has been developed.

Debye approximation is valid for the acoustic mode of vibrations, indicating that at lower values of frequency and wave vector, the dispersion curve can be approximated as a linear relation between ω and K, as shown in Fig. 3.17. K varies from 0 to $\frac{\pi}{a}$. The value of ω at which K is maximum and is equal to $\frac{\pi}{a}$ is called the Debye frequency.

The slopes of the optical modes of vibrations are relatively lower than those of the acoustic waves. Therefore, the optical phonons can be approximated efficiently by Einstein's approximation. Based on this, the plot of ω versus k is a horizontal line.

For small values of k, the linear relationship between ω and k can be written as

$$\omega = \sqrt{\frac{k}{m}} . \kappa a. \tag{3.40}$$

The unit of the term $\sqrt{\frac{k}{m}} a$ is m/sec. Therefore, the expression for ω can be written as

$$\omega = \upsilon_{Debye} \kappa, \tag{3.41}$$

where v_{Debye} is the Debye velocity. However, the Debye velocity is not exactly equal to the phonon velocity, which can be obtained by taking the slope of the dispersion curve.

The optical phonons can be approximated with a horizontal line (ω = constant). Therefore, the velocity of the optical phonons is 0. They are stagnant; however, they act like a capacitor, thus taking the heat and interacting with higher electromagnetic modes of vibration.

3.5 Density of States

Consider a 2-D crystal whose wave vector is quantized in both x- and y-directions. k_x and k_y represent the wave vectors in the x- and y-directions, respectively. The 2-D crystal can be graphically represented on an xy plot by plotting k_x vs. k_y. Discrete quantum states can be visualized by dividing this plot into equal segments. The wave vector for the crystal is $\frac{2\pi n}{L}$, where $n = \pm 1, \pm 2, \ldots$. The coordinate system can be established at a particular location of this diagram so that the length of each segment is equal to $\frac{2\pi}{L}$. The coordinates on the x-axis vary from $0, \frac{2\pi}{L}, \frac{4\pi}{L}$. Now, consider a differential element of a sphere, which can be indicated with a wave vector k having the width dk. Therefore, the total number of quantum states within the differential surface element spanned by the width dk can be obtained by dividing the total surface area of the differential elemental volume by the volume of each segment:

$$= \frac{4\pi k^2 dk}{V} = \frac{L^3 k^2 dk}{2\pi^2}. \tag{3.42}$$

The number density of quantum states accommodated per entire volume of a crystal is given by

$$\frac{k^2 dk}{2\pi^2}. \tag{3.43}$$

The density of states can be defined based on the number density of quantum states. Therefore, the density of states can be defined in terms of the number of states or number density of states per unit differential energy (Fig. 3.18):

$$D(E) = \frac{k^2}{2\pi^2} \frac{dk}{dE}. \tag{3.44}$$

Many microstates exist in a 3-D crystal structure. Therefore, it is difficult to establish a macroscale property by calculating the energy of each microstate. Hence, these discrete microstates should be approximated with a continuous representation. Therefore, the concept of density of states was developed to convert the discrete microstates into a continuous macroscale approximation.

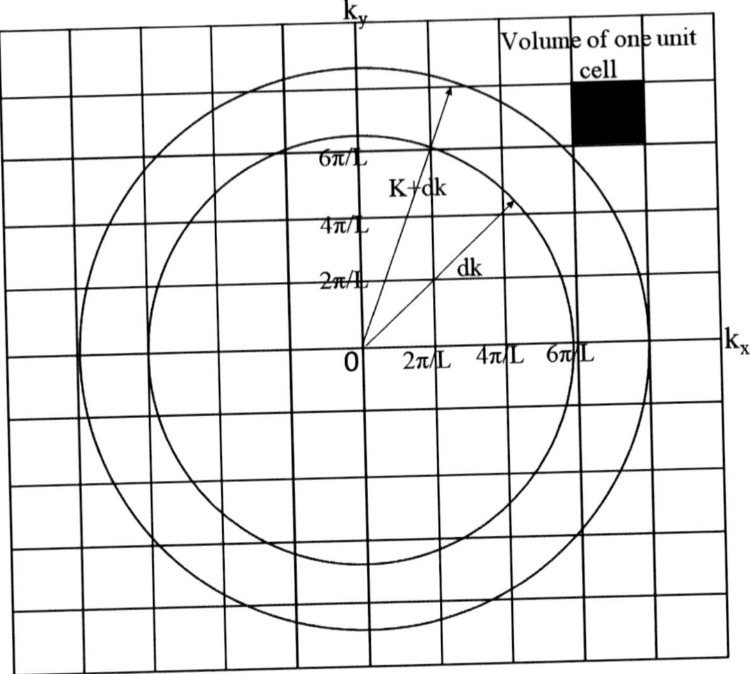

Fig. 3.18 Density of states

Therefore, from the expression for the density of states, it is clear that upon dividing by dE, the function for the discontinuous number density of quantum states is converted into a continuous function, $\frac{dk}{dE}$. The dispersion relation for metals is given by

$$E = E_c + \frac{\hbar^2 k^2}{2m^*},\tag{3.45}$$

where E_c is the energy corresponding to the conduction band and m^* is the effective mass of the crystal. We write the expression for k as

$$k = \sqrt{\frac{2m^*(E - E_c)}{\hbar^2}}.\tag{3.46}$$

Differentiating with respect to E,

$$\frac{dk}{dE} = \frac{1}{2}\left(\frac{2m^*}{\hbar^2}\right)^{1/2}\frac{1}{\sqrt{E - E_c}}.\tag{3.47}$$

Substituting with $\frac{dk}{dE}$ and multiplying with a factor 2, the expression for the density of states is given as

$$D(E) = \frac{1}{2\pi^2}\left(\frac{2m^*}{\hbar^2}\right)^{3/2}\sqrt{E - E_c}. \tag{3.48}$$

Note that the analysis of the electrons with a positive spin is being carried out. Therefore, the expression for the density of states should be multiplied by a factor '2' to obtain the total density of states, which accounts for the electron spin.

The density of states for electrons can be represented graphically by plotting $D(E)$ versus E (see Fig. 3.19). The graph shows that the plot starts from $E = E_c$ because the minima for the conduction band starts at E_c. Therefore, the density of states becomes zero at E_c, that is, the available quantum states are zero at E_c. From $E = E_c$, a parabolic variation can be observed. The quantum states start from the conduction band because the electrons start filling from lower energy levels to higher energy levels inside the conduction band. Therefore, the available quantum states will increase inside the conduction band. In general, the density of states is the number of quantum states that are available for the electrons to occupy.

The density of states is the number of states per unit volume per unit energy interval. Therefore, the number of quantum states per unit volume of the crystal can be obtained by integrating $D(E)$ with respect to E. Hence, the number of available quantum states (n) to be occupied from the beginning of the conduction band at $0\ K$ can be calculated by integrating $D(E)$ between the limits E_c and E_f:

$$n = \int_{E_c}^{E_f} D(E)dE. \tag{3.49}$$

The Debye approximation can be used to calculate the density of states for phonons:

$$\omega = v_D k. \tag{3.50}$$

Fig. 3.19 Graphical representation of the density of states for electrons

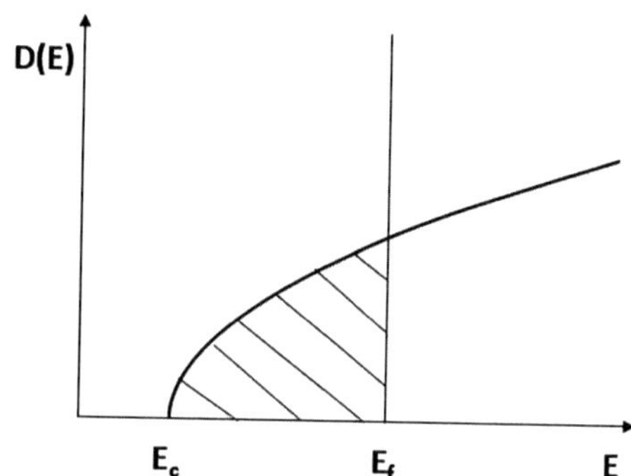

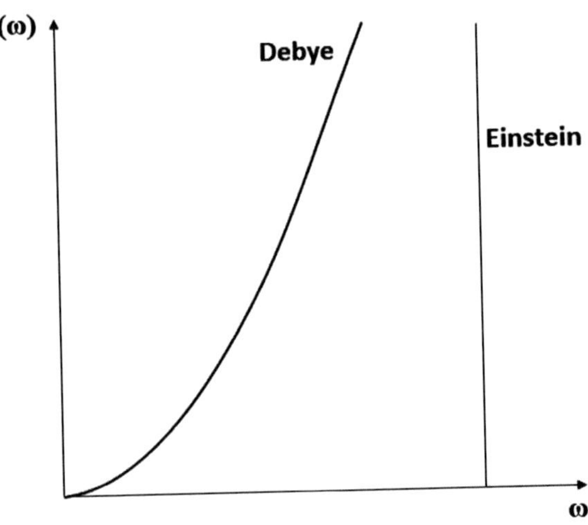

Fig. 3.20 $D(\omega)$ versus ω for acoustic phonons

For phonons, the frequency is used instead of energy. Therefore, the expression for the number of quantum states can be written as

$$n = \frac{1}{3\pi^2}\left(\frac{2m^*}{\hbar^2}\right)^{3/2}(E_f - E_c)^{3/2}. \tag{3.51}$$

There are three modes of vibration, including one longitudinal mode and two transverse modes, which exist in the case of phonons. Therefore, the density of states for phonons is given by

$$D(\omega) = \frac{4\pi k^2 dk}{\left(2\pi/L\right)^3}. \tag{3.52}$$

Note that the multiplication factor '3' corresponds to three modes of vibrations.

Therefore, $D(\omega)$ varies parabolically with ω for acoustic phonons, as shown in Fig. 3.20. For optical phonons, the plot of $D(\omega)$ versus ω is a vertical line. As stated earlier, the optical phonons follow the Einstein approximation.

3.6 Energy Quantization in Nanostructures

Consider a nanofilm, as shown in Fig. 3.21, of a width (d) of 10 nm in the Z direction that is not confined in the x and y coordinates. Therefore, the energy, which is a function of K_x, K_y, and K_z, becomes discrete only in the Z direction because of the confinement. Hence, the energy is quantized in the Z direction and will have the

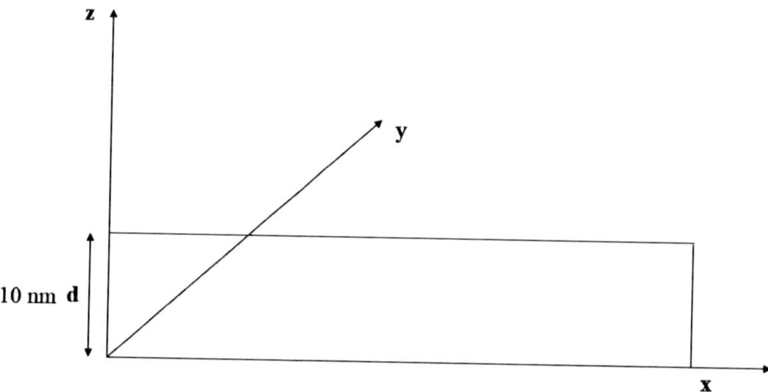

Fig. 3.21 Nanofilm

Fig. 3.22 Variation of $D(E)$ with E for electrons

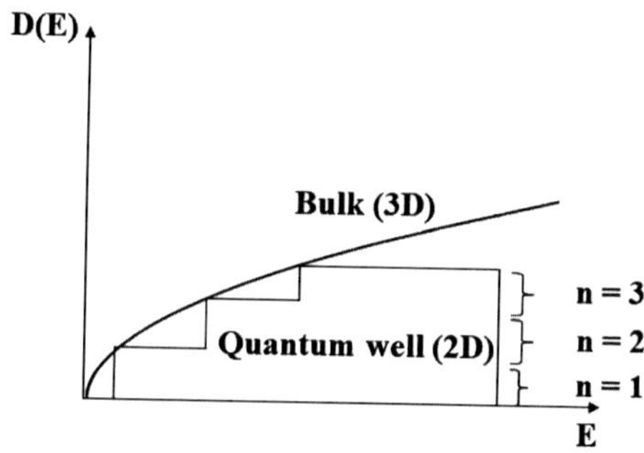

crystal wave vectors in the x- and y-directions. In the x- and y-directions, energy will have the form of the crystal wave vector. The energy is expressed as

$$E(k_x, k_y, k_z) = \frac{\hbar^2 k_{xy}^2}{m^*} + \frac{n^2 \hbar^2 \pi^2}{2m^* d^2}.$$
(3.53)

Therefore, the dispersion between E and K_{xy} is continuous, whereas it is discrete between E and K_z.

In the case of electrons in a 3-D crystal, the density of states $D(E)$ varies parabolically with E. For electrons confined in one direction, because of the quantization in the Z direction, a staircase profile can be observed corresponding to $n = 1, 2$, as shown in Fig. 3.22. Similarly, if the confinement is in two directions, discrete vertical lines called quantum wires can be seen. If the confinement is in all three directions, quantum dots can be observed. This is how the profile for the density of states varies based on the quantum confinement effects.

In this chapter, the analysis of the free electrons subjected to an arbitrary potential inside the crystal is carried out using the Kronig–Penney model. In addition, metals, insulators, conductors and semiconductors are classified based on the dispersion curves. The way to analyze the vibrational waves or phonons collectively for a large number of atoms is described using phonon dispersion. The concept of density of states is introduced to convert the discrete microstates into continuous macroscale approximations.

Exercise Problems

1. In the Kronig-Penney Model, for $P = 5$, find out the possible solutions for K_a. Convert the solution into a relationship between wave vector and energy and plot the solution in a reduced zone representation. You can write a program to aid the plotting process.
2. Consider a diatomic chain of atoms arranged periodically in 1-D. The masses of the two atoms are different but the spacing and the spring constant between them are the same. Derive the dispersion relationship between the phonon frequency and wave vector. Schematically draw the phonon dispersion you obtained.
3. Derive an expression for the electron density of states inside a quantum wire (constrained in 2-D and free in the third dimension). Plot this expression as a function of electron energy assuming a square cross section.
4. Assuming that phonons of a three-dimensional crystal obey the following isotropic dispersion relation, Derive an expression for the phonon density of states.

$$\omega = 2\sqrt{\frac{k}{m}} \left| \sin\left(\frac{ka}{2}\right) \right|$$

5. For the phonon dispersion, in the above problem, derive an expression for speed of sound in a solid. Calculate the speed of sound for a monatomic FCC crystal along (100) and (111) directions, using this expression. Assume that the mass of the atom is 9.31×10^{-23} kg, the lattice constant of the conventional FCC cell is 5.54 Å, and the spring constant is 7600 N/m.

Chapter 4
Fundamentals of Statistical Thermodynamics

4.1 Basic Concepts

Statistical thermodynamics is considered to be a bridge between microstates and macrostates. Simple particles in a quantum well or a potential constraint, electrons in a real crystal structure, and several quantum states in an energy carrier are some examples of microstates. In general, microstates correspond to the discontinuous wave vector. In the case of a particle in a quantum well, the wave vector $k = n\pi/D$, and in a real crystal, $k = 2\pi n/L$. Therefore, discontinuous values of k give rise to discontinuous values of energy. These discontinuous energy states are called microstates.

Internal energy is the property associated with macrostates. Therefore, the information related to microstates should be linked to macrostates in order to define the macrostate or macroscale property for a system with a large number of microstates. It is in this context that statistical thermodynamics becomes relevant.

In statistical thermodynamics, the fundamental building blocks of individual energy carriers at the discrete nature of the energy states are linked to the continuum picture to represent the microstates with macroscale variables. A pictorial representation of statistical thermodynamics is shown in Fig. 4.1.

4.1.1 Distribution Functions

Electrons or phonons in energy states cannot be fulfilled randomly. They occupy the energy states in a particular manner. The distribution of electrons or phonons in energy states can be denoted by distribution functions. Therefore, statistical thermodynamics provides the order in which the energy states are filled by energy carriers and transfers this information from microscale to macroscale. The distribution of energy carriers in the energy states can be elucidated further with a simple example.

© The Author(s) 2025

A. Pattamatta and S. K. Das, *Fundamentals of Nano- and Microscale Heat Transport*,
https://doi.org/10.1007/978-3-031-89613-2_4

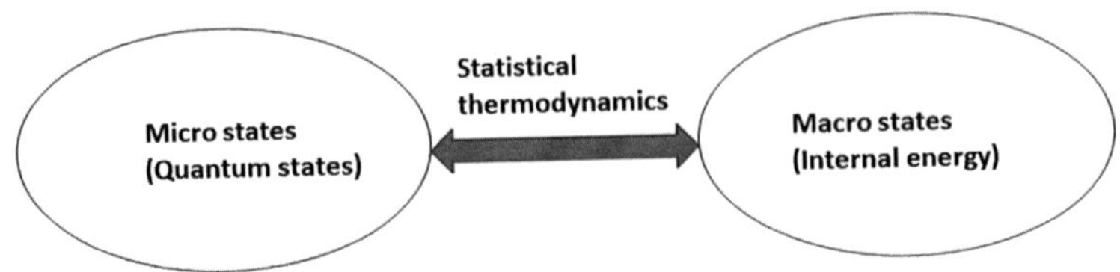

Fig. 4.1 Statistical thermodynamics

Consider three distinguishable particles and four microstates $\epsilon_0, \epsilon_1, \epsilon_2, \epsilon_3$, where ϵ_0 refers to the ground state. These three distinguishable particles can occupy the energy states in three different scenarios. In the first scenario, A, all the three particles can occupy the state ϵ_1 and the other states are entirely vacant. In the second scenario, B, two particles can occupy ϵ_0 and the third particle can occupy the highest energy state ϵ_0. The middle states ϵ_1 and ϵ_3 are entirely vacant. In the third scenario, C, the first particle occupies ϵ_0, the second particle occupies ϵ_2, the third particle occupies ϵ_3, and ϵ_4 is vacant.

The possible number of quantum states in each of the above scenarios is given as follows:

Scenario A:

$$\Omega(a) = \frac{3!}{0!3!0!0!} = 1. \tag{4.1}$$

Scenario B:

$$\Omega(b) = \frac{3!}{2!0!0!1!} = 3. \tag{4.2}$$

Scenario C:

$$\Omega(c) = \frac{3!}{1!1!1!0!} = 6. \tag{4.3}$$

The maximum probable distribution corresponds to the most significant value of Ω.

In general, if there are n_i number of particles distributed in an energy state ϵ_i, then the maximum possible number of microstates is given by

$$\Omega = \frac{n!}{n_0!n_1!n_1!, \ldots, n_r!}, \tag{4.4}$$

where r is the maximum number of energy levels. The most probable distribution can be obtained by maximizing Ω.

Applying natural log on both sides, we obtain

$$\ln \Omega = \ln n! - \ln \left(\prod_{i=0}^{r} n_i! \right). \tag{4.5}$$

For a large number of x_i values, the Sterling approximation can be used:

$$\ln X! = X \ln X - X. \tag{4.6}$$

Applying the Sterling approximation, the equation becomes

$$\ln \Omega = (n \ln n - n) - \sum_{i=1}^{r} (n_i \ln n_i - n_i). \tag{4.7}$$

Consider a closed system consisting of n number of particles. They can occupy all the energy states under few constraints. The first constraint is that the total number of particles (n) in a closed system remains constant. Therefore,

$$n = \text{constant} \tag{4.8}$$

$$\Rightarrow n_0 + n_1 + n_2 + \cdots + n_r = n \tag{4.9}$$

$$\Rightarrow \sum_{i=0}^{r} n_i = n = \text{constant}. \tag{4.10}$$

Therefore, the change in n should be equal to 0, as shown in the following equation:

$$\delta n = 0. \tag{4.11}$$

The second constraint is that the total internal energy should also be equal to a constant, i.e., 5

$$U = \text{constant}. \tag{4.12}$$

Since the given system is closed, there is no transfer of energy to or from the surroundings. Therefore, the total internal energy remains constant:

$$\Rightarrow n_0 \epsilon_0 + n_1 \epsilon_1 + n_2 \epsilon_2 + \cdots + n_r \epsilon_r = \text{constant} \tag{4.13}$$

$$\Rightarrow \sum_{i=0}^{r} n_i \epsilon_i = \text{constant}. \tag{4.14}$$

Therefore,

$$\delta U = 0. \tag{4.15}$$

For the most probable distribution that tries to maximize Ω, even after interchanging some particles in the energy levels, the value of Ω should not change considerably because it is already close to the maximum probability. Therefore, we can assume that at the maximum probable distribution, even a small rearrangement of particles should not affect the value of Ω. Therefore,

$$\delta \ln \Omega = 0. \tag{4.16}$$

Applying δ to Eq. (4.7), we obtain

$$-\sum \left(\delta n_i \ln n_i + n_i \frac{\delta n_i}{n_i} - \delta n_i \right) = 0 \tag{4.17}$$

$$\Rightarrow \sum \delta n_i \ln n_i = 0. \tag{4.18}$$

This equation provides the distribution of particles, which maximizes the probability Ω. It also satisfies the condition that the total number of particles cannot change during the rearrangement of particles in the energy states. From this perspective, the concept of Lagrange multipliers becomes relevant. A constant β, which has the units of reciprocal energy, can be introduced. Upon applying the Lagrange multiplier β to Eq. (4.14), we obtain

$$\sum \beta \epsilon_i \delta n_i = 0, \tag{4.19}$$

where the constant β can be written inside the summation.

Similarly, another Lagrange multiplier α can be introduced. Multiplying Eq. (4.10) with α, we obtain

$$\sum \alpha \delta n_i = 0, \tag{4.20}$$

where α is a constant having no units. Therefore, upon adding Eqs. (4.18) to (4.20), we observe

$$\sum_{i=0}^{r} (\ln n_i + \alpha + \beta \epsilon_i) \delta n_i = 0. \tag{4.21}$$

Expanding this equation,

$$\Rightarrow (\ln n_0 + \alpha + \beta \epsilon_0) \delta n_0 + (\ln n_1 + \alpha + \beta \epsilon_1) \delta n_1 + \cdots = 0, \tag{4.22}$$

where δn_0 and δn_1 cannot be equal to 0 because there is an absolute change in the distribution. But,

$$\delta n_0 + \delta n_1 + \delta n_2 + \delta n_3 + \cdots = 0. \tag{4.23}$$

Therefore,

$$\ln n_i + \alpha + \beta \epsilon_i = 0. \tag{4.24}$$

Hence, the most probable distribution that maximizes the value of ω is given by

$$P(E_i) = \frac{n_i}{\sum\limits_{i=0}^{r} n_i} \Rightarrow n_i = \frac{1}{e^\alpha e^{\beta \epsilon_i}} = e^{-(\alpha + \beta \epsilon_i)}. \tag{4.25}$$

This equation satisfies both the constraints: $n = $ constant and $U = $ constant.

4.1.2 *Microcanonical, Canonical, and Grand-Canonical Ensembles*

When ϵ increases, the number density of particles in the energy levels decreases because a number of particles occupy the lower energy states and taper off at the higher energy states. Therefore, various kinds of systems, such as microcanonical, canonical, and grand-canonical ensembles, should be introduced. Table 4.1 shows a comparison of these systems.

The ensemble is a statistical agglomeration of the distribution functions and is used to define the macroscale properties. In the microcanonical ensemble, the distribution will have equal probabilities.

The probability distribution can be written as

$$P_i = \frac{1}{\Omega}. \tag{4.26}$$

This ensemble can be applied to an isolated system in which the internal energy, volume, and the number of particles are constant. In canonical and grand-canonical ensembles, the probability distribution can be obtained by maximizing Ω. The canonical ensemble can be applied to an isolated system in which the volume, temperature, and the number of particles are fixed.

The major difference between canonical and grand-canonical ensembles is that the latter is applicable to an open system. Therefore, the number of particles "n" is not a constant. Volume (V) and chemical potential (μ) are constant. Also, the grand-canonical ensemble can be applied to an isothermal system in which the temperature is constant.

In both canonical and grand-canonical ensembles, the probability distribution can be calculated by

$$P(E_i) = \frac{n_i}{\sum\limits_{i=1}^{r} n_i}. \tag{4.27}$$

Table 4.1 Comparison of microcanonical, canonical, and grand-canonical ensembles

Microcanonical	Canonical	Grand-canonical
$P_i = \frac{1}{\Omega}$	$P(E_i) = \dfrac{n_i}{\sum\limits_{i=1}^{r} n_i}$	$P(E_i) = \dfrac{n_i}{\sum\limits_{i=1}^{r} n_i}$
Can be applied to an isolated system	Can be applied to an closed system	Can be applied to an open system
U, V, N are fixed	V, N, T are fixed	V, μ are fixed
	$\alpha = 0$ $\beta = \frac{1}{k_B T}$	$\alpha = -\frac{\mu N_i}{k_B T}$ $\beta = \frac{1}{k_B T}$

However, the values of the Lagrange multipliers are different. In canonical ensembles, $\alpha = 0$ and $\beta = \frac{1}{k_B T}$. In grand-canonical ensemble, $\alpha = \frac{-\mu N_i}{k_B T}$ and $\beta = \frac{1}{k_B T}$. Therefore, the distribution function can be written in terms of probability distribution as follows:

$$P(E_i) = e^{\frac{-E_i}{k_B T}}. \tag{4.28}$$

For grand-canonical ensembles,

$$P(E_i) = \frac{e^{-(E_i - \mu N_i)/k_B T}}{\sum\limits_{i=0}^{r} e^{-(E_i - \mu N_i)/k_B T}}. \tag{4.29}$$

While the canonical ensembles can be applied to a system of molecules, phonons, and photons, the grand-canonical ensembles can be applied to a system of electrons.

In the probability distribution for the canonical ensemble, the numerator $e^{\frac{-E_i}{k_B T}}$ is called the Boltzmann distribution and the denominator is considered to be the partition function, which is the summation of distributions across different energy levels. Sometimes, Z can be used instead of writing the summation.

4.2 Equilibrium Distribution for Different Statistics: Maxwell–Boltzmann, Fermi–Dirac, Bose–Einstein Functions

On the basis of the equilibrium distribution functions, the individual distribution functions for each energy carrier can be derived. For example, in the case of a monoatomic gas molecule that follows the Maxwell–Boltzmann distribution, only translational kinetic energy exists, which is equal to

$$E = \frac{1}{2} m (v_x^2 + v_y^2 + v_z^2). \tag{4.30}$$

Similarly, for phonons,

$$E_i = h\upsilon \left(n + \frac{1}{2} \right). \tag{4.31}$$

While phonons and photons follow the Bose–Einstein distribution function, electrons follow the Fermi–Dirac distribution function.

To calculate the probability distribution, partition functions should be evaluated. Appropriate values of energies corresponding to molecules, phonons, and photons should be substituted into the generic expression for partition functions.

4.2.1 *Maxwell–Boltzmann Distribution Function*

In the case of monoatomic molecules, the translational kinetic energy is given by

$$E = \frac{1}{2}m(v_x^2 + v_y^2 + v_z^2), \tag{4.32}$$

where v_x, v_y, and v_z are the velocities in the x-, y-, z-directions, respectively. The generic expression for the partition function is given as

$$Z = \sum e^{\frac{-E_i}{k_B T}}. \tag{4.33}$$

Molecules belong to the canonical ensemble. Accordingly, the partition functions are to be evaluated. In continuous functions, integral can be used instead of summation. Z can be written as

$$Z = \int\limits_{-\infty}^{\infty} \int\limits_{-\infty}^{\infty} \int\limits_{-\infty}^{\infty} e^{\frac{-m}{2}\left(v_x^2 + v_y^2 + v_z^2\right)/k_B T}\, dv_x dv_y dv_z. \tag{4.34}$$

Solving the expression, the partition function (Z) can be obtained as

$$Z = \left(\frac{m}{2\pi k_B T}\right)^{-3/2}. \tag{4.35}$$

The probability distribution function $P(E_i)$ is given by

$$P(E_i) = \left(e^{\frac{-E_i}{k_B T}}\right)\left(\frac{m}{2\pi k_B T}\right)^{3/2}. \tag{4.36}$$

The number of molecules that can occupy an energy level has to be evaluated.

Therefore, the number distribution can be evaluated by multiplying the probability distribution with the number density. Number density (n) is the number of molecules per unit volume of a system:

$$f = P(E_i)n \Rightarrow f = n\left(e^{-\frac{m(v_x^2 + v_y^2 + v_z^2)}{2k_B T}}\right)\left(\frac{m}{2\pi k_B T}\right)^{3/2}. \tag{4.37}$$

This distribution function is called the Maxwell–Boltzmann distribution function. It can be called classical distribution function because the distribution of energy in the energy levels is continuous. Therefore, principles of quantum mechanics are not needed to calculate the distribution functions.

4.2.2 *Fermi–Dirac Distribution Function*

For the electrons that obey the grand-canonical ensemble, the probability distribution function is given by

$$P(E_i) = \frac{e^{-(E_i - \mu N_i)/k_B T}}{Z}. \tag{4.38}$$

The partition function is the summation of the numerator and is equal to

$$Z = \sum e^{-(E_i - \mu N_i)/k_B T}. \tag{4.39}$$

The empty energy levels can be represented by $N_i = 0$; therefore, energy $E_i = 0$ since there are no electrons in the energy levels. The filled energy level can be represented by $N_i = 1$. Therefore, the partition function is the summation that runs over both empty and filled energy levels.

$$Z = e^{-0} + e^{-(E_i - \mu)/k_B T}$$
$$\Rightarrow Z = 1 + e^{-(E_i - \mu)/k_B T}. \tag{4.40}$$

The probability distribution function is given by

$$P(E_i) = \frac{e^{-(E_i - \mu N_i)/k_B T}}{1 + e^{-(E_i - \mu)/k_B T}}. \tag{4.41}$$

Therefore, the number density function (f) can be written as

$$f = n_i P(E_i) \Rightarrow 0 \times P(E_i) + 1 \times P(E_i)$$
$$\Rightarrow f = \frac{e^{-(E_i - \mu)/k_B T}}{1 + e^{-(E_i - \mu)/k_B T}} \tag{4.42}$$
$$\Rightarrow f = \frac{1}{1 + e^{(E_i - \mu)/k_B T}}.$$

This distribution function (f) is called the Fermi–Dirac distribution function. Note that the number density function is derived based on only filled and unfilled energy states. However, for phonons, the number density function can be derived based on all the possible discrete energy states. The Fermi–Dirac distribution function can be represented graphically by plotting the number density function corresponding to the energy state E_i on the Y-axis and E_i on the X-axis for various temperatures (T), as shown in Fig. 4.2.

If

$$E - \mu \geq k_B T \Rightarrow 1 + e^{-(E_i - \mu)/k_B T} = e^{-(E_i - \mu)/k_B T}, \tag{4.43}$$

then this becomes the Boltzmann distribution function. Therefore, for the limiting case, the Fermi–Dirac distribution reduces to the Boltzmann distribution function.

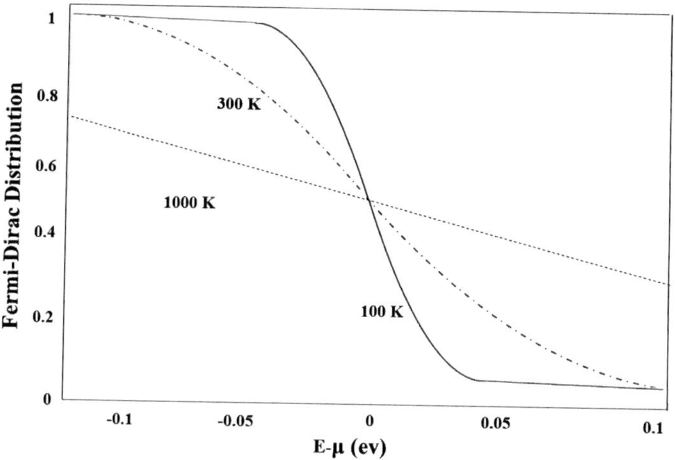

Fig. 4.2 Physical representation of the Fermi-Dirac distribution function for various temperatures

4.2.3 Bose–Einstein Distribution Function

For phonons, the vibrational energy Z can be obtained by substituting the vibrational energy of phonons as follows:

$$E = h\upsilon \left(n + \frac{1}{2} \right), \tag{4.44}$$

where $n = 0, 1, 2, 3, \ldots$ Therefore, the partition function Z can be written as

$$Z = \sum_{n=0}^{\infty} e^{-h\upsilon(n+\frac{1}{2})/k_B T} \Rightarrow \sum_{n=0}^{\infty} e^{-h\upsilon/2k_B T} e^{-nh\upsilon/k_B T}. \tag{4.45}$$

Writing the term $e^{\frac{-h\upsilon}{2k_B T}}$ outside and expanding the summation, we obtain

$$Z = e^{\frac{-h\upsilon}{2k_B T}} \left(1 + e^{\frac{-h\upsilon}{k_B T}} + \left(e^{\frac{-h\upsilon}{k_B T}} \right)^2 + \cdots \right). \tag{4.46}$$

The Maclaurin series expansion of $(1 - x)^{-1}$ is given by

$$(1 - x)^{-1} = 1 + x + x^2 + x^3 + \cdots \tag{4.47}$$

Therefore,

$$Z = e^{\frac{-h\upsilon}{2k_B T}} \left(1 - e^{\frac{-h\upsilon}{k_B T}} \right)^{-1}. \tag{4.48}$$

The probability distribution function $P(E_i)$ can be written as

$$P(E_i) = \frac{e^{-h\upsilon(n+\frac{1}{2})/k_B T}}{e^{\frac{-h\upsilon}{2k_B T}}\left(1 - e^{\frac{-h\upsilon}{k_B T}}\right)^{-1}}, \tag{4.49}$$

$$P(E_i) = e^{\frac{-nh\upsilon}{k_B T}}\left(1 - e^{\frac{-h\upsilon}{k_B T}}\right). \tag{4.50}$$

Therefore, the number density function (f) can be written as

$$f = \langle n \rangle = \sum_{n=0}^{\infty} n_i P(E_i)$$

$$\Rightarrow \sum_{n=0}^{\infty} e^{\frac{-nh\upsilon}{k_B T}} n \left(1 - e^{\frac{-h\upsilon}{k_B T}}\right) \tag{4.51}$$

$$\Rightarrow \left(1 - e^{\frac{-h\upsilon}{k_B T}}\right) \sum_{n=0}^{\infty} n e^{\frac{-nh\upsilon}{k_B T}}.$$

Applying the differentiation rule, we obtain

$$-\frac{d}{dx} \sum_{n=0}^{\infty} e^{-nx} = \sum_{n=0}^{\infty} n e^{-nx}. \tag{4.52}$$

Therefore, the number density function becomes

$$f = \left(1 - e^{\frac{-h\upsilon}{k_B T}}\right)\left(-\frac{d}{dx} \sum_{n=0}^{\infty} e^{\frac{-nh\upsilon}{k_B T}}\right). \tag{4.53}$$

Substituting

$$\frac{h\upsilon}{k_B T} = x, \tag{4.54}$$

the number density function becomes

$$f = \frac{\left(1 - e^{-x}\right)e^{-x}}{\left(1 - e^{-x}\right)^2} \Rightarrow \frac{1}{e^x - 1}, \tag{4.55}$$

which is called the "Bose–Einstein distribution function."

The Bose–Einstein distribution function can be represented physically by plotting f on the Y-axis and frequency (υ) on the X-axis for various temperature values, as shown in Fig. 4.3. If $h\upsilon \geq k_B T$, then the number density function becomes

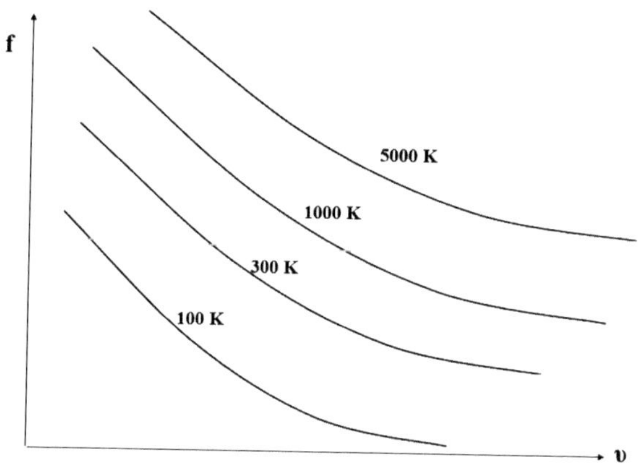

Fig. 4.3 Physical representation of the Bose–Einstein distribution function for various temperatures

$$f = e^{\frac{-h\upsilon}{k_B T}}.$$
(4.56)

Therefore, for the limiting case, the number density function (f) approaches the Boltzmann distribution. The expression for the density of states for electrons is given by

$$D(E) = \frac{1}{V}\frac{dk}{dE}.$$
(4.57)

The number of quantum states per unit volume of crystal per unit differential energy is called the density of states.

The total number of quantum states (N) per unit volume of the entire crystal can be obtained by integrating $D(E)$:

$$\frac{N}{V} = \int_0^\infty D(E)dE.$$
(4.58)

The number density or distribution function (f) can be defined as the number of energy carriers occupying a particular energy level per unit volume. Therefore, integrating the product of f and $D(E)$ over the entire energy spectrum gives the total number of energy carriers or particles distributed across the quantum states. With this information, the macroscopic property of internal energy can be calculated. In this way, the statistical thermodynamics is connecting microstates to macrostates.

Assume that a box is filled with several monotonic gas molecules. The translational kinetic energy possessed by each gas molecule is given as follows:

$$\frac{1}{2}m\left(v_x^2 + v_y^2 + v_z^2\right).$$
(4.59)

The generic expression for the internal energy per unit volume for the entire system can be obtained by the following equation:

$$U = \sum E_i f_i, \tag{4.60}$$

where f_i is the distribution function corresponding to the energy carrier. Therefore, the internal energy per unit volume for the entire system can be obtained by using

$$U = \int\limits_{-\infty}^{\infty} \int\limits_{-\infty}^{\infty} \int\limits_{-\infty}^{\infty} \frac{-m}{2} \left(v_x^2 + v_y^2 + v_z^2 \right) \left(\frac{m}{2\pi k_B T} \right)^{3/2} dv_x dv_y dv_z. \tag{4.61}$$

The Maxwell–Boltzmann distribution function is used because the energy carriers are molecules.

Upon evaluating the integral, the internal energy is expressed as follows:

$$U = \frac{3}{2} n k_B T. \tag{4.62}$$

Specific heat capacity (C_v) can be obtained by differentiating volumetric internal energy (U) with respect to temperature (T) as follows:

$$C_v = \frac{dU}{dt} = \frac{3}{2} n k_B \Rightarrow C_v = \frac{3}{2} R_u. \tag{4.63}$$

The term $n k_B$ is called the universal gas constant (R_u). The real gas constant can be obtained by dividing R_u with molecular mass.

The readers should be aware of a very famous theorem in statistical thermodynamics called the equipartition theorem. Equipartition theorem states that the internal energy contribution in each direction is equal to $\frac{1}{2} k_B T$. Therefore, the overall internal energy contribution for all the three directions is equal to $\frac{3}{2} k_B T$. Each degree of freedom contributes to $\frac{1}{2} k_B T$ amount of energy. On plotting the variation of $\frac{C_v}{R_u}$ versus $\log T$ for a diatomic ideal gas, as shown in Fig. 4.4, approximately up to 80 K, $\frac{C_v}{R_u}$ is equal to 1.5. $\frac{C_v}{R_u}$ is called the normalized heat capacity. As the temperature increases, the contributions from the rotational motion become very significant and affect the normalized heat capacity. In simple words, the $\frac{C_v}{R_u}$ value from the translational degrees of freedom is 3/2. On adding another degree of freedom due to rotation, $\frac{C_v}{R_u}$ becomes 5/2. On further increasing the temperature beyond 1000 K, its value becomes 7/2 because the vibrational mode is also taken into account.

In the case of phonons, the Bose–Einstein distribution function is used to calculate the internal energy of solids. In the case of acoustic phonons, according to the Debye approximation, the expression for the density of states is given by

$$D(\omega) = \frac{3\omega^2}{2\pi^2 v_D^3}, \tag{4.64}$$

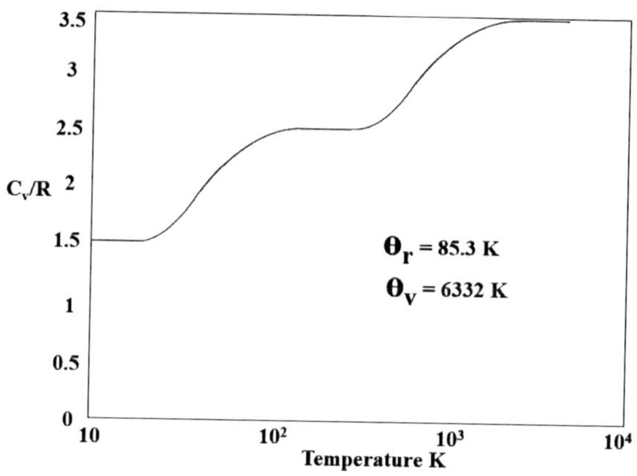

Fig. 4.4 Variation of $\frac{C_v}{R_u}$ versus log T

where $\omega = v_D k$ and V_D is the Debye velocity. This is only valid for acoustic phonons. The optical phonons follow the Einstein approximation. Therefore, $\frac{dk}{d\omega}$ becomes infinity.

4.3 Internal Energy and Specific Heat

The internal energy is expressed by

$$U = \sum E_i f_i. \tag{4.65}$$

The energy (E_i) of one quantum state is $\hbar\omega$. Therefore, internal energy is the product of $\hbar\omega$ and distribution function, which in turn is a function of ω and T

$$U = \sum \hbar\omega.f(T,\omega). \tag{4.66}$$

Therefore, the expression for density of states should be multiplied by internal energy to convert this summation into a continuous integral.

Therefore, the expression for internal energy becomes

$$U = \int_0^{\omega_D} \hbar\omega.f(T,\omega)D(\omega)d\omega. \tag{4.67}$$

In the case of acoustic phonons under the Debye approximation, the minimum and maximum limits for the density of states are 0 and ω_D because the Brillouin zone varies from 0 to ω_D. Therefore, the final expression for internal energy by substituting the Bose–Einstein distribution function is given as

$$U = \frac{3}{2\pi^2 v_D^3} \int_0^{\omega_D} \frac{\hbar\omega^3}{e^{\hbar\omega/k_B T} - 1} d\omega. \tag{4.68}$$

The volumetric heat capacity can be obtained by differentiating internal energy with respect to T as follows:

$$C_v = \frac{dU}{dT} = \frac{3\hbar^2}{2\pi^2 v_D^3 k_B T^2} \int_0^{\omega_D} \frac{\omega^4 e^{\hbar\omega/k_B T}}{e^{\hbar\omega/k_B T} - 1} d\omega. \tag{4.69}$$

The number of quantum states per unit volume of the crystal can be obtained by integrating the density of states from the limits 0 to ω_D as follows:

$$\frac{N}{V} = \int_0^{\omega_D} D(\omega) d\omega = \frac{\omega_D^3}{2\pi^2 v_D^3}. \tag{4.70}$$

Since the total number of acoustic phonons are 3, the expression for the number of quantum states per unit volume of the crystal can be rewritten, as shown in the following equation:

$$\frac{N}{V} = 3 \int_0^{\omega_D} D(\omega) d\omega = 3\frac{\omega_D^3}{2\pi^2 v_D^3}. \tag{4.71}$$

Rewriting the expression for specific heat in terms of the number of quantum states per unit volume, we obtain

$$C_v = 9k_B \left(\frac{N}{V}\right) \left(\frac{T}{\theta_D}\right)^3 \int_0^{\frac{\theta_D}{T}} \frac{x^4 e^x}{(e^x - 1)^2} dx, \tag{4.72}$$

where $x = \frac{\hbar\omega}{k_B T}$ and the Debye temperature $\theta_D = \frac{\hbar\omega}{k_B}$. Note that the new limits of integration vary from 0 to $\frac{\theta_D}{T}$.

Electrons are present in semiconductors, dielectrics, and metals. In the dispersion curve, the Fermi level for metals exists inside the conduction band, as shown in Fig. 4.5. Therefore, calculation of the heat capacity in metals is slightly different from that in semiconductors.

Volumetric internal energy is the summation of the product of quantum state per unit volume and the energy associated with that quantum state. Therefore, the volumetric internal energy of electrons per quantum state can be calculated as follows:

$$U = \sum E f_{fD}(E, T, \mu), \tag{4.73}$$

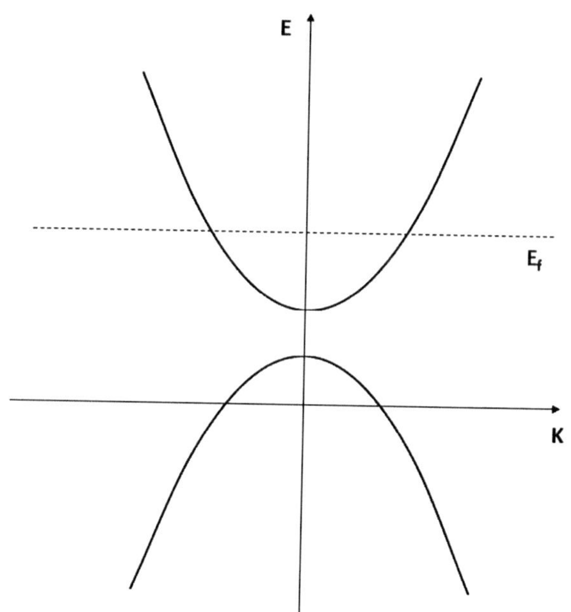

Fig. 4.5 Dispersion curve for metals

where f_{fD} is the Fermi–Dirac distribution function, which is a function of electron energy (E), temperature (T), and chemical potential (μ). The chemical potential can also be taken as the Fermi energy level E_f.

The internal energy for all the quantum states in the entire crystal can be calculated by multiplying the expression for volumetric internal energy with a density of states $D(E)$ as follows:

$$U = \int_0^\infty E f_{fD}(E, T, \mu) D(E) dE. \tag{4.74}$$

Therefore, the discrete summation is converted into a continuous integral. The limits of integration vary from 0 to ∞ because the electrons start from the beginning of the conduction band and they can move freely up to ∞.

The number of electrons per unit volume can be calculated by integrating the product of the density of states and distribution function as follows:

$$n_e = \int_0^\infty f_{fD}(E, T, \mu) D(E) dE. \tag{4.75}$$

By adding and subtracting

$$\int_0^\infty E_f f(E, T, \mu) D(E) dE$$

to the expression for internal energy, we obtain

$$U = \int_0^\infty (E - E_f) f(E, T, \mu) D(E) dE + \int_0^\infty E_f f(E, T, \mu) D(E) dE. \qquad (4.76)$$

Fermi level (E_f) is the chemical potential at $0\ K$. Chemical potential is defined as the potential up to which the electrons can be filled into the energy levels.

The volumetric heat capacity for electrons can be obtained by

$$C_e = \frac{dU}{dt}. \qquad (4.77)$$

The second term on the right-hand side is 0 because it can be written as $E_f n_e$, which is a constant. The derivative of a constant is 0. Therefore, the expression for volumetric heat capacity becomes

$$C_e = \int_0^\infty (E - E_f) \frac{df}{dT} D(E) dE. \qquad (4.78)$$

$\frac{df}{dT}$ can be evaluated by differentiating the Fermi–Dirac distribution function with respect to T.

The Fermi–Dirac distribution function (f) is given by

$$f = \frac{1}{e^{\left(\frac{E-\mu}{k_B T}\right)} + 1}. \qquad (4.79)$$

Therefore,

$$\frac{df}{dT} = \frac{(E - \mu)}{k_B T^2} \frac{e^{\left(\frac{E-\mu}{k_B T}\right)}}{\left(e^{\left(\frac{E-\mu}{k_B T}\right)} + 1\right)^2}. \qquad (4.80)$$

Substituting $\frac{df}{dT}$, the expression for specific heat is given as

$$C_e = \int_0^\infty (E - E_f) \frac{(E - \mu)}{k_B T^2} \frac{e^{\left(\frac{E-\mu}{k_B T}\right)}}{\left(e^{\left(\frac{E-\mu}{k_B T}\right)} + 1\right)^2} D(E) dE. \qquad (4.81)$$

Upon plotting the Fermi–Dirac distribution function (f_{fD}) as a function of E at 0 K, one can observe that the occupancy function is 1 up to the Fermi energy level ($E = E_f$). Therefore, at 0 K, the distribution function obeys the Dirac-delta function. This is due to the fact that the electrons occupy up to the Fermi level, above which there are no free electrons.

At a temperature of 100 K, the distribution function shifts from Dirac-delta function to a continuous function. At very high temperatures, it gradually changes and becomes an inclined line because, at very high temperatures, the electrons can freely move to the conduction band. Therefore, there is a fair chance that the energy levels above and below the Fermi levels can also be occupied.

The value of $\frac{df}{dT}$ does not change considerably in the energy levels that are typically close to the lowest and highest energy levels. Primarily, the maximum change in $\frac{df}{dT}$ occurs about the Fermi level. Therefore, around the Fermi level, the density of states is nearly a constant. Hence, the density of states near the Fermi level ($D(E_f)$) can be taken out of integration.

Therefore, rewriting the equation for volumetric specific heat (C_e) around the Fermi level by considering a differential element of width dE_f, we obtain

$$C_e = D(E_f) = \int_{E_f - dE_f}^{E_f + dE_f} \frac{\left(E - E_f\right)^2}{k_B T^2} \frac{e^{\frac{E - E_f}{k_B T}}}{\left(e^{\frac{E - E_f}{k_B T}} + 1\right)^2} dE. \tag{4.82}$$

Note that the chemical potential (μ) is replaced with E_f and the limits of integration around the Fermi level become $E - dE_f$ and $E + dE_f$, respectively.

By substituting $\frac{E - E_f}{k_B T} = x$ in the original equation for specific heat, we obtain

$$C_e = k_B^2 T D(E_f) \int_{\frac{-E_f}{k_B T}}^{\infty} \frac{x^2 e^x}{(e^x + 1)^2} dx = \frac{\pi^2}{3}. \tag{4.83}$$

The number of electrons is calculated by

$$n_e = \int_0^{\infty} f^c D(E). \tag{4.84}$$

At 0 K,

$$n_e = \frac{2}{3} E_f D(E_f). \tag{4.85}$$

Therefore, substituting n_e, the expression for specific heat is obtained as follows:

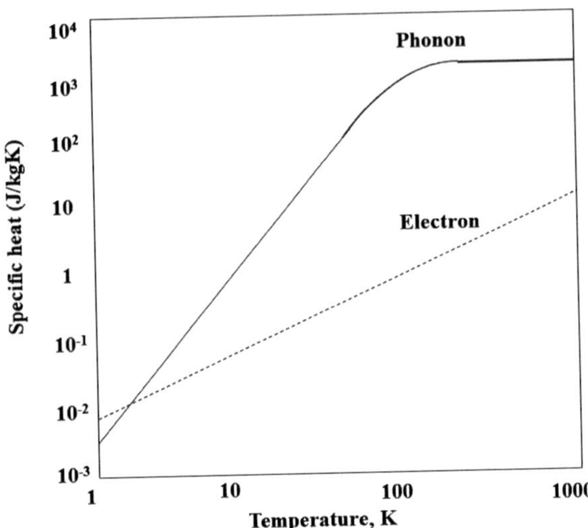

Fig. 4.6 Variation of C_v with temperature (T)

$$C_e = \frac{1}{2}\pi^2 n_e k_B \frac{T}{T_f}, \qquad (4.86)$$

where the Fermi temperature $T_f = \frac{E_f}{k_B}$.

The variation of C_v with temperature (T) can be shown graphically by plotting C_v on the Y-axis and $\log T$ on the X-axis, as shown in Fig. 4.6. In the case of electrons, C_v varies linearly with T. For phonons, C_v is given by

$$C_v = 9k_B \left(\frac{N}{V}\right)\left(\frac{T}{\theta_D}\right)^3 \int_0^{\frac{\theta_D}{T}} \frac{x^4 e^x}{(e^x - 1)^2}dx. \qquad (4.87)$$

At very low temperatures, the lower limit $\frac{\theta_D}{T}$ tends to ∞. On evaluating the integral, C_v is obtained as a function of the cubic power of T (C_v is proportional to T^3.) At elevated temperatures, $\frac{\theta_D}{T}$ tends to 0. C_v becomes a constant equivalent to $\frac{9N}{Vk_B}$. Therefore, for lower temperatures, the graph is slightly curved and for elevated temperatures, the plot shows a horizontal line. For a moderate temperature range, linear variation can be observed.

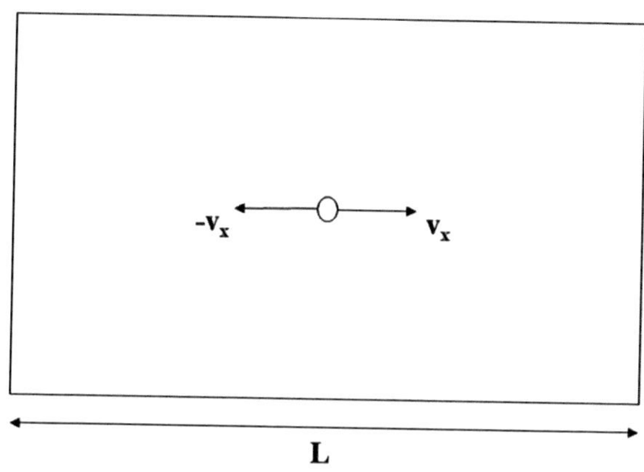

Fig. 4.7 Motion of a gas molecule in a rectangular container

4.4 Kinetic Theory

4.4.1 Estimation of Pressure, Temperature, and Mean Free Path

Consider a simplified model based on kinetic theory, in which a rectangular container contains gas molecules, as shown in Fig. 4.7. The length of the container is L along the Cartesian x-direction. Assume that the motion of gas molecules is translational only in either positive or negative x-direction. The velocity in the positive direction is v_x and that in the negative direction is $-v_x$. The kinetic energy of the gas molecules increases with temperature.

According to the Newtonian physics, the impulse experienced by the gas molecule on the right wall, at $x = 1$, is given by

$$F\Delta t = mv_x - (mv_x) = 2mv_x, \tag{4.88}$$

where Δt is the time between two successive collisions, which are responsible for the change in momentum. The gas molecules hit the wall with a positive momentum mv_x and rebound back with $-mv_x$. Therefore, the change in momentum is $mv_x - (-mv_x) = 2mv_x$. Assume that the collision is perfectly elastic. Hence, the gas molecules rebound with the same momentum in the opposite direction. The distance traveled by the gas molecules between two successive collisions is $2L$. Therefore, the velocity possessed by the gas molecules is $\frac{2L}{\Delta t}$.

Substitution of the velocity in the impulse expression provides

$$F = \frac{mv_x^2}{L}. \tag{4.89}$$

This is the expression for the force exerted by the gas molecule on the right wall due to the motion of the molecules in the x-direction. The analysis can be extended to all the three directions by replacing v_x with the absolute velocity v. Therefore,

$$v^2 = v_x^2 + v_y^2 + v_z^2.$$ (4.90)

Assume that $v_x = v_y = v_z$. Therefore,

$$v^2 = 3v_x^2 \Rightarrow v_x^2 = \frac{v^2}{3}.$$ (4.91)

This expression for force can be written as

$$F = \frac{mv^2}{3L}.$$ (4.92)

The pressure exerted by the gas molecules due to the collision is given as

$$P = \frac{F}{A} = \frac{mv^2}{3AL}.$$ (4.93)

For N number of molecules, the total pressure exerted by the gas molecules is given by

$$P = \frac{N}{V} \frac{mv^2}{3}.$$ (4.94)

The volume of the container $(V) = AL$, and the number density $(\hat{n}) = \frac{N}{V}$. Therefore, the expression for pressure becomes

$$P = \hat{n} \frac{mv^2}{3}.$$ (4.95)

The pressure can be expressed in terms of kinetic energy as follows:

$$P = \frac{2}{3} \hat{n} KE.$$ (4.96)

In the case of the Maxwell distribution of molecules, the translational kinetic energy is equal to the internal energy. Therefore, the expression for kinetic energy is given by

$$U = KE = \frac{3}{2} \hat{n} KE.$$ (4.97)

By substituting $\frac{3}{2}k_B T = \frac{1}{2}mv^2$ in the expression for pressure, we obtain

$$PV = Nk_B T.$$ (4.98)

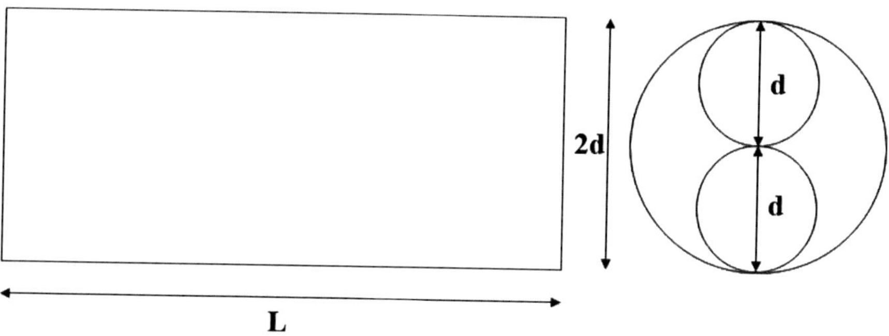

Fig. 4.8 Collision of two molecules in a container

Dividing and multiplying by the number of moles (n), we obtain

$$PV = n \left(\frac{N}{n} \right) k_B T, \tag{4.99}$$

where $\frac{N}{n}$ is the Avogadro number (N_A). Thus,

$$PV = n N_A k_B T \Rightarrow PV = n R_u T. \tag{4.100}$$

Here,

$$N_A k_B = R_u, \tag{4.101}$$

where R_u is the universal gas constant.

Upon multiplying and dividing with molecular weight (m_w), we obtain

$$PV = n m_w \bar{R} T, \tag{4.102}$$

where the mass of the gas (m) = $n m_w$. Molecular weight can be expressed in terms of kg/kmol.

Therefore, according to kinetic theory, both macroscale properties, pressure and temperature, are only functions of kinetic energy. The molecular mean free path can be obtained by using kinetic theory.

Assume that two molecules of diameter d collided with each other in a cylinder of length L, as shown in Fig. 4.8. The number density of the molecules within this volume is $\hat{n}$. The total number of molecules that can collide within this volume can be obtained by multiplying the number density and volume, which is given by $\hat{n} \pi d^2 L$.

Therefore, effectively, the molecule must travel a distance of mean free path (Λ) to collide with another molecule. Hence, the expression becomes

$$\hat{n} \pi d^2 \Lambda = 1. \tag{4.103}$$

Therefore, the expression for mean free path (Λ) is given by

$$\Lambda = \frac{1}{\hat{n}\,\pi\,d^2}. \tag{4.104}$$

The expression for density is given by

$$\rho = \frac{M}{V} = \frac{n\,M_w}{V} \Rightarrow \frac{N}{N_A}\frac{M_w}{V}, \tag{4.105}$$

where $\frac{N}{V} = n$ and

$$\hat{n} = \frac{\rho N_A}{M_w}. \tag{4.106}$$

Therefore, the expression for mean free path becomes

$$\Lambda = \frac{M_w}{\pi d^2 N_A \rho}, \tag{4.107}$$

where $\frac{R}{N_A} = k_B$.

Therefore,

$$\Lambda = \frac{k_B T}{\pi d^2 P}. \tag{4.108}$$

It is assumed that the system of two collided molecules moves and collides with that of the other molecule. However, the other molecule is already in motion. Therefore, to account for that correction, the effective mean free path has to be reduced because both the molecules are in motion. Hence, the expression for the mean free path has to be divided with a correction factor of $\sqrt{2}$:

$$\Lambda = \frac{1}{\sqrt{2}}\frac{k_B T}{\pi d^2 P}. \tag{4.109}$$

The mean free path can be calculated for the gas molecules at room temperature by substituting the following values in the expression for the mean free path: $T = 300\ K$; $P = 1$ atm $= 10.5\,\mathrm{N/m^2}$; the effective diameter (d) for the gas molecules $= 2.5 \times 10^{-10}$ m; and the Boltzmann constant (k_B) $= 3.8 \times 10^{-23}$ J/K. The expression for mean free path (Λ) then becomes 149 nm.

4.4.2   Estimation of Thermal Conductivity

To solve the problems associated with heat transport, apart from the properties such as specific or volumetric heat capacities, another fundamental thermophysical property

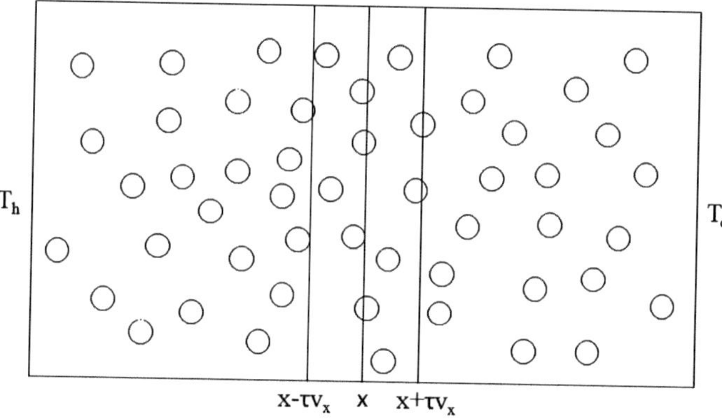

Fig. 4.9 Heat conduction phenomenon in a container

called thermal conductivity has to be estimated. Evaluation of the thermal conductivity of various materials is difficult. The size has a significant role on the estimation of thermal conductivity. The thermal conductivity changes can be observed clearly in the nanoscale materials. Therefore, thermal conductivity can be understood fundamentally by nonequilibrium heat transport phenomenon. One of the simplest ways of analyzing the thermal conductivity is by using the kinetic theory.

Some elementary kinetic theory concepts can be applied to obtain the expressions for properties, such as pressure, temperature, mean free path, thermal conductivity, and viscosity.

The application of this kinetic theory in the case of heat conduction can be illustrated as follows. Consider a container having randomly dispersed ideal gas molecules, as shown in Fig. 4.9. The left wall of the container is maintained at a higher temperature T_h and the right wall is maintained at a lower temperature (T_c). The top and bottom walls are insulated, such that the heat transfer takes place effectively only from left to right.

The kinetic theory can be applied to calculate the heat flux for this system. In the first step, a vertical line can be drawn at the middle portion of the container. This vertical line can be treated as an imaginary interface, which allows the flux of the particle density to move from one end to the other. Some particles move from the left wall to the right wall, and some move from the right wall to the left wall. However, effective flux moves from the left wall to the right wall. The particle number density across the interface is $\hat{n}$. Let the coordinate of the interface be x. Two vertical lines are drawn about the interface, such that the coordinates of these two lines are $x - \Lambda$ and $x + \Lambda$, respectively, where Λ is the molecular mean free path. To relate the mean free path (Λ) with velocity, another parameter called the relaxation time (τ) is introduced. Relaxation time is the timescale between two successive collisions. The velocity of the gas molecules is v_x. The mean free path can be related to the relaxation time as $v_x \tau$. Therefore, the relaxation time converts the length scales into timescales and vice versa.

Therefore, the particle travels a distance of $x + \Lambda$, $x - \Lambda$ before it collides with the next particle in the positive and negative x-directions, respectively. Therefore, there is an equal probability that half of the molecules travel toward the right and another half toward the left before colliding with the other molecules. There is an equal probability that 50% of the molecules move toward the hot side and 50% of the molecules move toward the cold side of the interface. However, the net heat flux direction is from the left end to the right end. Therefore, the net heat flux across the interface at x can be obtained by subtracting half of the number density of molecules located at $x + \tau v_x$ from half of the number density of molecules at $x - \tau v_x$ as follows:

$$Q_x = \frac{1}{2} \left(\hat{n} \, E \, v_x \right)\Big|_{x-\tau v_x} - \frac{1}{2} \left(\hat{n} \, E \, v_x \right)\Big|_{x+\tau v_x}. \tag{4.110}$$

From the Taylor series expansion, the energy flux at the location $x + \tau v_x$ can be written as

$$\left(\hat{n} \, E \, v_x \right)\Big|_{x+\tau v_x} = \left(\hat{n} \, E \, v_x \right)\Big|_{x} + \frac{\partial}{\partial x} \left(\left(\hat{n} \, E \, v_x \right) \right) \tau v_x. \tag{4.111}$$

Similarly, the energy flux at $x - \tau v_x$ is given by

$$\left(\hat{n} \, E \, v_x \right)\Big|_{x-\tau v_x} = \left(\hat{n} \, E \, v_x \right)\Big|_{x} - \frac{\partial}{\partial x} \left(\left(\hat{n} \, E \, v_x \right) \right) \tau v_x. \tag{4.112}$$

Therefore, the net heat flux can be obtained by subtracting the heat flux at $x + \tau v_x$ from the heat flux at $x - \tau v_x$ as follows:

$$Q_x = -\frac{\partial}{\partial x} \left(\hat{n} \, E \, v_x^2 \right) \tau = -\tau v_x^2 \frac{dU}{dx}, \tag{4.113}$$

where $\hat{n} \, E = U$. Internal energy U is a function of temperature. Therefore,

$$Q_x = -\tau v_x^2 \frac{dU}{dT} \frac{\partial T}{\partial x}, \tag{4.114}$$

where $\frac{dU}{dT} = C_v$ (volumetric heat capacity). Hence, the final expression for heat flux is given by

$$Q_x = -\tau v_x^2 C_v \frac{dT}{dx}. \tag{4.115}$$

Comparing this equation with Fourier heat conduction law, we obtain

$$Q_x = -k \frac{dT}{dx}, \tag{4.116}$$

where $k = \tau v_x^2 C_v$ and $v_x^2 = \frac{v^2}{3}$.

Therefore, the expression for k becomes

$$k = \frac{\tau v^2 C}{3}.$$
(4.117)

The total thermal conductivity of a semiconductor device is the summation of the individual thermal conductivities of both phonons and electrons. Therefore, we obtain

$$k = k_c + k_p.$$
(4.118)

In semiconductor devices, electron contribution may be less compared to that of phonons. But in the calculation of the thermal conductivity, these two components are taken into account. Analogously, the expression for molecular viscosity can be obtained from the kinetic theory.

4.4.3 Estimation of Viscosity

Using Couette flow, the expression for viscosity can be obtained. Consider two parallel plates: the bottom plate is fixed and the top plate is of infinite length moving at velocity u, as shown in Fig. 4.10. Therefore, the velocity variation is primarily in the y-direction. The separation distance between the plates is h. The velocity profile is linear in the vertical direction. The maximum velocity (u) can be observed at the top plate. The velocity variation can be obtained by

$$u(y) = \frac{uy}{h}.$$
(4.119)

To obtain the expression for thermal conductivity, the net flux of energy for heat conduction is calculated. In flow problems, the flow is governed by momentum gradient. Therefore, the net flux of momentum is taken into account. A line can be

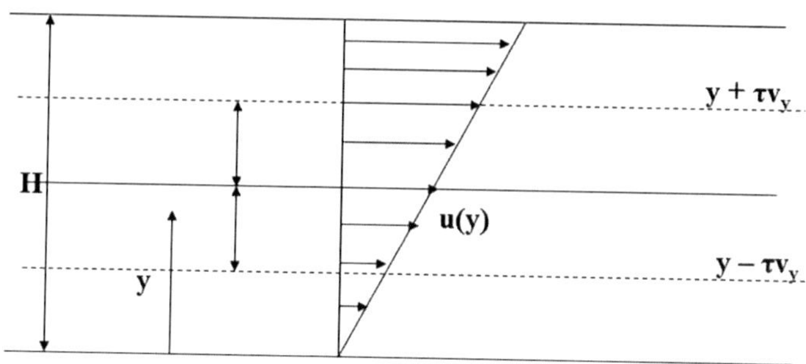

Fig. 4.10 Couette flow

drawn in the horizontal direction at a distance of y from the bottom plate to represent the interface.

Top and bottom boundaries are drawn in such a way that they are at a distance of the mean free path from the interface. Therefore, the coordinates of the top and bottom faces are $y + \tau v_y$ and $y - \tau v_y$, respectively. The net flux density of momentum in the y-direction is equal to shear stress (τ), which can be calculated as follows:

$$\tau = \left(\frac{1}{2} \frac{\hat{n}}{N_A} m_w v_x v_y \right)_{y - \tau v_y} - \left(\frac{1}{2} \frac{\hat{n}}{N_A} m_w v_x v_y \right)_{y + \tau v_y} . \tag{4.120}$$

Compared to Newton's law of viscosity,

$$\tau = \mu \frac{\partial u}{\partial y} . \tag{4.121}$$

The expression for viscosity (μ) is given by

$$\mu = \frac{1}{3} \rho v \Lambda , \tag{4.122}$$

where $\Lambda = \frac{1}{\sqrt{2} n \pi d^2}$.

Substituting Λ, the expression for viscosity, we obtain

$$\mu = \frac{m v}{3 \sqrt{2} \pi d^2} . \tag{4.123}$$

In this chapter, the concepts of the distribution function, microcanonical, canonical, grand-canonical ensembles are clearly illustrated in order to understand the distribution of electrons and phonons in energy states. The expressions for pressure, temperature, mean free path, thermal conductivity, and viscosity are derived using the kinetic theory.

Exercise Problems

1. A semiconductor can be approximated with a parabolic band structure for the dispersion relation. The Fermi level in the semiconductor could be above or below the conduction band edge. Assume the effective mass as the free electron mass. For $|\mu - E_c| = 0.05$ eV and $T = 300$ K, do the following in the range 0 eV $< E - E_c < 0.1$ eV:

   (a) Plot the Fermi-Dirac distribution as a function of E.
   (b) Plot the density of states as a function of E.

(c) Calculate the product of $f(E, T)D(E)$, which means the average number of electrons at each allowable energy level E, and plot the product as a function of E.

(d) Calculate the product $Ef(E, T)D(E)$, which means the actual energy at each allowable energy level E, and plot the product as a function of E.

2. Plot the Bose–Einstein distribution as a function of frequency for $T = 100$ K, 300 K, and 1000 K. Compare with the Boltzmann distribution at the same temperatures.

3. A crystal has a Debye velocity of 5000 m/s, and a Debye temperature of 500 K. For $T = 300$ K,

(a) Plot the Bose–Einstein distribution as a function of ω.

(b) Plot the density of states as a function w using the Debye model.

(c) Plot $f \cdot D(\omega)$ as a function of ω.

(d) Plot $\omega \cdot D(\omega.)$ as a function of ω.

(e) Compute the specific heat of the crystal as a function of temperature for $1 < T < 1000$ K.

4. Derive the expression for the electron internal energy and specific heat for a quantum well nanostructure, as a function of temperature. Assume that the electron effective mass is equal to the free electron mass and the electron number density $n_e = 2 \times 10^{28}$ m^{-3}.

5. Assuming that the phonons obey the sinusoidal dispersion relation derived in class, derive an expression for the phonon internal energy and specific heat.

6. The Debye temperature of diamond is 1320 K. Calculate the specific heat of diamond at 300 K and compare it with the literature value. The lattice constant of diamond is 3.567 Å.

Chapter 5
Nanoscale Transport Processes

5.1 Particle Transport Process

5.1.1 Quantum States

The knowledge of the distribution function is significant for the study of transport phenomena. Many quantum states exist in a system, and each quantum state will have a number of distributions of n_i, as shown in Fig. 5.1, where $i = 1, 2, 3, \ldots$ The last quantum state can be represented by n_Ω. Therefore, the summation of n_i is equal to the total number of particles (N) distributed across the quantum states:

$$\sum n_i = N. \tag{5.1}$$

Each of the quantum states can be represented graphically by a dot in the plot of the position vector ($\vec{r}$) on the horizontal coordinate and the momentum ($\vec{p}$) vector on the vertical coordinate, respectively, as shown in Fig. 5.2.

In a plot of the quantum states or microstates from 1 to Ω, discrete points or dots can be visualized. The ensemble of all these microstates will form a macrostate. Each quantum state moves almost with the same velocity and position. The ensemble of these particles travels with respect to time due to advection. Therefore, there is change in both position and momentum with time. Hence, the relative motion and momentum with respect to time must be studied along with the average energy possessed by the ensemble of these particles, which gives information about the transport process. A plot of position and momentum vectors is called a phase diagram.

If there are m dimensions and N particles in a system, then the total number of degrees of freedom (n) with respect to position is given by

$$n = mN. \tag{5.2}$$

© The Author(s) 2025

A. Pattamatta and S. K. Das, *Fundamentals of Nano- and Microscale Heat Transport,*
https://doi.org/10.1007/978-3-031-89613-2_5

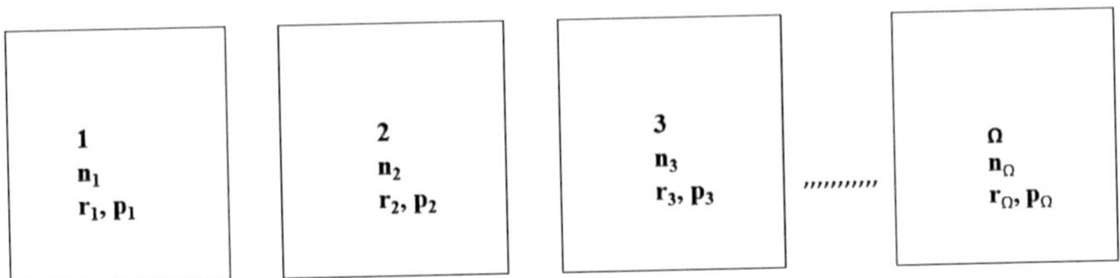

Fig. 5.1 Discrete quantum states

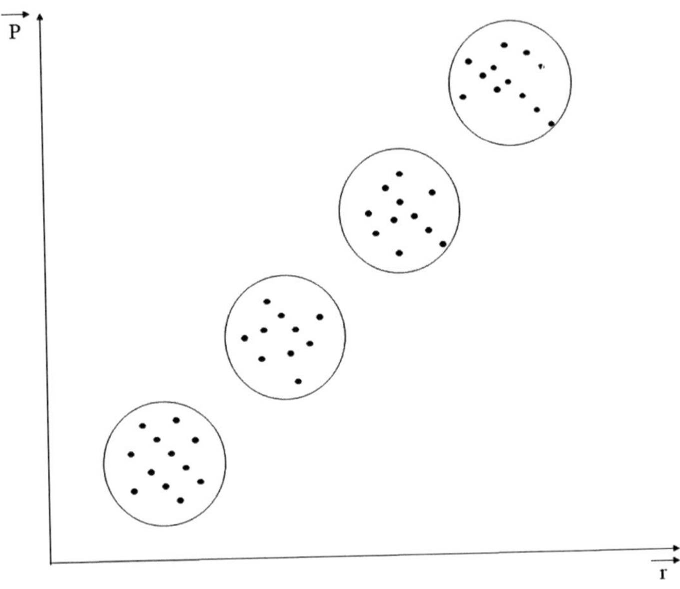

Fig. 5.2 Representation of a quantum state

Considering the momentum, the total number of degrees of freedom is given as

$$n = 2mN.$$ (5.3)

Therefore, for 3-D space, the total number of degrees of freedom is given by

$$n = 2.3.N = 6N.$$ (5.4)

5.1.2 Liouville Equation and Boltzmann Transport Equation (BTE)

Consider an elemental control volume in the pr space having the dimensions dr and dp. According to the conservation principle, the net rate of particles or points flowing into the control volume minus the rate of particles or points flowing out of the control volume is equal to the rate of change in the number of particles inside the control volume.

This conservation principle can be expressed in a mathematical form. Let f be the particle number distribution function. In transport phenomena, f is the nonequilibrium distribution function, whereas in statistical thermodynamics, it is the equilibrium distribution function. Therefore, f is a function of both position (r) and momentum (p).

The rate of change in particle distribution within the control volume for a system of N number of particles is $\left(\frac{\partial f}{\partial t}\right)^{(N)}$.

The rate of change in the distribution function f with respect to position (r) for N number of particles is $\left(\frac{\partial f}{\partial r}\right)^{(N)}$. This must advect with a velocity of $\frac{dr}{dt}$ or $\dot{r}$, where r is a vector. Therefore, the conservation term with respect to position in a tensorial form for n degrees of freedom is given as

$$\sum_{i=1}^{n} r_i \times \left(\frac{\partial f}{\partial r_i}\right)^{(N)}. \tag{5.5}$$

Similarly, the conservation term with respect to momentum can be written as

$$\sum_{i=1}^{n} p_i \times \left(\frac{\partial f}{\partial p_i}\right)^{(N)}. \tag{5.6}$$

Note that the analysis cannot be done for a single particle. An ensemble of particles should be considered. The ensemble with N number of particles will have n degrees of freedom. Therefore, the advection terms have to be summed up over the number of degrees of freedom and should be collectively represented with respect to both position and momentum. Therefore, the conservation equation can be expressed as

$$\left(\frac{\partial f}{\partial t}\right)^{(N)} + \sum_{i=1}^{n} r_i \times \left(\frac{\partial f}{\partial r_i}\right)^{(N)} + \sum_{i=1}^{n} p_i \times \left(\frac{\partial f}{\partial p_i}\right)^{(N)} = 0. \tag{5.7}$$

This equation is also called the Liouville equation.

Therefore, the Liouville equation holds for a system with N number of particles. However, it is cumbersome to track the distribution with N number of particles. In an N-particle system, the process of understanding the distribution function for each microstate and then summing them over all the degrees of freedom is computationally expensive. Therefore, reducing the equation to a single-particle distribution function is a more practical approach. A drastic reduction can be observed in the number of degrees of freedom. Therefore, by reducing the N-particle distribution to a one-particle distribution, the Liouville equation becomes

$$\frac{\partial f}{\partial t} + \frac{d\vec{r}}{dt} \cdot \nabla_r f + \frac{d\vec{p}}{dt} \cdot \nabla_p f = \left(\frac{\partial f}{\partial t} \right)_c. \tag{5.8}$$

This equation is called the Boltzmann transport equation.

While converting an N-particle system to a single-particle system, an additional term called the collision term is added, which accounts for the collision between the particles. With the reduction in the degrees of freedom, the complexity caused by advection is reduced, but a new term called collision term is introduced to satisfy the conservation equation.

Here, momentum $p = mv$, $\frac{dr}{dt} = \vec{v}$, and $\frac{dp}{dt} = \vec{f}$. Substituting these into Eq. (5.8), the Boltzmann transport equation becomes

$$\frac{\partial f}{\partial t} + \vec{v}\, \nabla_r f + \frac{\vec{f}}{m} \cdot \nabla_v f = \left(\frac{\partial f}{\partial t} \right)_c. \tag{5.9}$$

From quantum mechanics, the momentum can be expressed as

$$p = \hbar k. \tag{5.10}$$

Therefore, substituting this relation into the Boltzmann transport equation,

$$\frac{\partial f}{\partial t} + \vec{v}\, \nabla_r f + \frac{\vec{f}}{\hbar} \cdot \nabla_k f = \left(\frac{\partial f}{\partial t} \right)_c. \tag{5.11}$$

This equation contains three spatial coordinates, three momentum coordinates, and one time coordinate. Therefore, it can be considered to be a 7-D problem. Computational complexity can be reduced drastically by converting the N-particle system to a single-particle system.

The collision term can be explained with a simple example of a collision between two particles, as shown in Fig. 5.3. The coordinates of the first particle are r, k, and t, where r, k, and t represent the position, wave vector, and momentum, respectively. The other particle is also at the same position r but with a different wave vector k_1. The coordinates of the second particle are r, k_1, and t. Assume that the collision is inelastic, in which both momentum and energy changes can be observed after the collision. The coordinates of the particles after the collision are r, k^1, t and r, k_1^1, t.

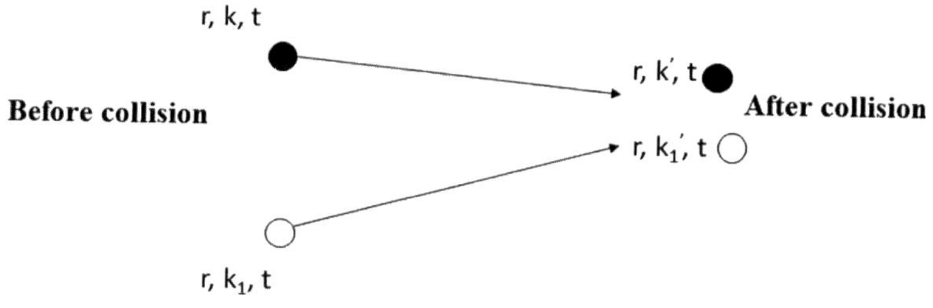

Fig. 5.3 The collision between two particles

Therefore, the collision term can be written as

$$\left(\frac{\partial f}{\partial t}\right)_c = -\int f(r,k,t)f(r,k_1,t)W(k,k_1) \to W(k^1,k_1^1)d^3k_1d^3k^1d^3k_1^1$$

$$+ \int f(r,k^1,t)f(r,k_1^1,t)W(k^1,k_1^1) \to W(k,k_1)d^3k_1d^3k^1d^3k_1^1.$$

$$(5.12)$$

The first term represents the collision term before collision and the second term represents the collision term after collision. Therefore, the rate of change in scattering can be obtained by subtracting the before-collision term from the after-collision term.

Before collision, a certain number of particles with wave vectors k, k_1 changed to k^1, k_1^1. After collision, particles with wave vectors k^1, k_1^1 changed to k, k_1. The W function in the collision term is called the Wronskian function, which signifies the relationship between changes in the wave vector from an initial wave vector due to collision.

The integrals contained in the collision term are called the scattering integrals. There are nine integrals in the collision term.

5.1.3 Relaxation Time Approximation to BTE

Solving the scattering integrals is computationally very expensive because the wave vectors before and after the collision for each particle need to be identified. To reduce the computational cost, researchers applied relaxation time approximation.

Based on relaxation time approximation (RTE), a very simple and practically useful one, the scattering or collision term can be expressed as

$$\left(\frac{\partial f}{\partial t}\right)_c = \frac{f - f_{eq}}{\tau},$$

$$(5.13)$$

where f is the equilibrium distribution function and τ is the relaxation time, which has the units of time.

The inspiration for relaxation time approximation comes from the Bhatnagar–Gross–Krook (BGK) approximation, which is the most widely used model in rarefied gas dynamics. Although the Boltzmann transportation equation is used to solve problems in rarefied gas dynamics, owing to the collision term, there exists computational complexity. Therefore, BGK approximation is used to reduce the computational cost.

Therefore, by applying the BGK approximation to the right-hand side of the Boltzmann transport equation and ignoring the advection and force terms, the equation becomes

$$\frac{\partial f}{\partial t} = \frac{f - f_{eq}}{\tau}. \tag{5.14}$$

Solving the equation, we obtain

$$f - f_{eq} = Ce^{\frac{-t}{\tau}}. \tag{5.15}$$

This is the solution in the case where the distribution function is not a function of space and momentum but only a function of time. When the time t approaches infinity, $f - f_{eq}$ tends to 0. Therefore, the nonequilibrium distribution function is approaching equilibrium distribution. The equilibrium distribution function is uniform in space. Therefore, on plotting the equilibrium distribution function with respect to momentum space or physical coordinate space, it appears like a circle. However, the nonequilibrium function appears to be distorted. Therefore, the relaxation time is the timescale over which relaxation from nonequilibrium distribution to equilibrium distribution takes place.

This can happen only through scattering or collision. This is why the entire collision term is incorporated only through the scattering time or the relaxation time. At higher collision rates, the relaxation times are smaller because the particles are tightly packed.

This is the practical approximation that is commonly used to solve the transport problems. By implementing BGK or relaxation time approximation, the Boltzmann transport equation becomes

$$\frac{\partial f}{\partial t} + \vec{v}\,\nabla_r f = \frac{f - f_{eq}}{\tau}. \tag{5.16}$$

Consider a case without any external force, and for the sake of non-dimensionalization, reference velocity u_{ref} can be substituted into the actual dimensional velocity. Similarly, a reference characteristic length scale L can be chosen.

The non-dimensional parameters are given by $x^* = \frac{x}{L}$, $y^* = \frac{y}{L}$. Therefore, the non-dimensional Boltzmann transport equation can be written as

$$\frac{\partial f}{\partial t^*} + \vec{v}^* \nabla_{r^*} f = \frac{f - f_{eq}}{Kn}. \tag{5.17}$$

In this equation, $\tau u_{ref} = \Lambda$, where Λ is the mean free path and $\frac{\Lambda}{L} = \text{Kn}$, where Kn is the Knudsen number. When the Knudsen number tends to 0, the right-hand side term becomes infinity.

In the continuum regime, the advection process is very insignificant compared to the collision process. Therefore, all the macroscopic laws of continuum, such as Fourier's equation, Newton's shear stress law, Ohm's law, and Fick's law of diffusion, can be derived under these conditions. On the contrary, when the Knudsen number approaches infinity, the right-hand side term in the BTE becomes 0. This is called ballistic transport in which there is no collision between the energy carriers. Consider two boundaries: one is maintained at a higher temperature (T_h) and the other is maintained at a lower temperature (T_c). If an energy carrier travels from the hot side to the cold side, it does not have information on the temperature of the cold side because there are no energy carriers in between the boundaries. Therefore, solving this problem using the Boltzmann transport equation, a temperature jump at the boundaries can be observed. When the Knudsen number approaches 0, the temperature profile varies linearly because a diffusion process dominates the energy transport and a classical conduction profile based on Fourier's equation can be observed. This analysis shows that the Boltzmann transport equation is a self-sufficient equation that carries all the information concerning the operating regime. Therefore, the Boltzmann transport equation is valid for all Knudsen numbers ranging from 0 to ∞.

5.1.4 Phonon Scattering

The information about several modes of scattering can be related to relaxation time. The scattering phenomena can be divided into several modes such as electron–electron scattering, electron–phonon scattering, phonon–phonon scattering, phonon–impurity scattering, and phonon–boundary scattering. Phonon transport is most dominant in semiconductor devices. Hence, phonon–phonon scattering takes place. Also, other impurities or dislocations exist in semiconductor devices. These impurities or dislocations in the crystal cause a resistance to the flow of lattice vibrations. Therefore, while modeling semiconductor devices, scattering of phonons along with dislocations or impurities should be considered. This is why phonon–impurity scattering came into existence.

At high Knudsen numbers, when phonons travel from left to right, assume that there are no impurities and no other phonons. Hence, there are no phonon–phonon scattering and phonon—impurity scattering. Therefore, the collision term becomes 0. However, phonon—boundary scattering takes place because phonons cannot propagate through the boundary, and they get scattered from the boundary surface. In most metals and semiconductors, electron–phonon scattering is significant. The free electrons can move in the crystal, colliding with the phonons. Therefore, a finite value of electrical resistance exists. If there is no electron–phonon collision, the electrical resistance is 0 and the electrons can freely flow through the metals or semiconductors. Hence, the electrical resistance can be explained by electron–phonon

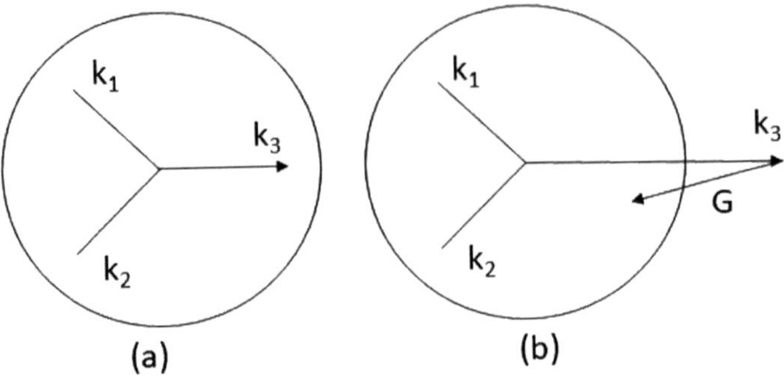

Fig. 5.4 a Normal scattering; **b** umklapp scattering

scattering. Similarly, the thermal resistance can be explained by phonon–phonon scattering. Electron–phonon scattering is a highly inelastic scattering process. The energy and wave vector change after collision. The electrons cannot interact with all kinds of phonons. Most likely, they interact with high-frequency phonons and optical phonons, which releases electromagnetic energy.

Consider a system of two phonons. The wave vectors of these phonons are k_1 and k_2, respectively. These two phonons collide with each other and a third phonon is formed with a wave vector k_3. The boundary in which these phonons collide with each other can be assumed to be a Brillouin zone. In this zone, the periodic repetition of the unit lattice structure takes place. In the case of a monoatomic lattice, the minimum allowable wavelength is $2a$. Hence,

$$\Lambda \geq 2a.$$

Therefore, after the collision, the wave vector of the third phonon should not be higher than the allowable wave vector space within the Brillouin zone. If the wave vector exceeds the Brillouin zone wave vector, then it is supposed to move out of the Brillouin zone. However, this is not possible because the vibrational wave vector cannot be larger than the lattice spacing. Therefore, the wave vector has to be corrected by pushing it back into the Brillouin zone and this can be done by introducing a reciprocal wave vector. Therefore, scattering in which the vibrational wave vector does not exceed the Brillouin zone is called normal scattering and scattering in which the wave vector should be corrected back using the reciprocal wave vector is called umklapp scattering. Normal and umklapp scatterings are shown in Fig. 5.4.

Writing the conservation of momentum and energy for the normal phonon scattering, we obtain

$$k_1 + k_2 = k_3. \tag{5.18}$$

For umklapp scattering,

$$k_1 + k_2 = k_3 - G. \tag{5.19}$$

The reciprocal wave vector is 0 in normal scattering, but it takes a specific value in umklapp scattering.

In other words, in the case of the two-phonon scattering process, since the edge of the Brillouin zone is $\frac{\pi}{a}$,

$$k_1 + k_2 \geq \frac{\pi}{a}. \tag{5.20}$$

This can be corrected by the reciprocal lattice vector and brought inside the Brillouin zone because this is the maximum allowable value of k, corresponding to the minimum wavelength (Λ). In umklapp scattering, the reciprocal lattice vector causes the resistance to momentum transfer. However, in normal scattering, there is no resistance to the collision of phonons. Hence, the momentum is brought down by using the reciprocal lattice vector, which results in thermal resistance.

Thermal conductivity of a solid is finite due to umklapp scattering. If there was no umklapp scattering, thermal conductivity would have been infinite because phonons can flow smoothly without any resistance. Therefore, the umklapp scattering process contributes to the finite thermal conductivity of solids.

The expression for relaxation time in umklapp scattering is given by

$$\tau_u^{-1} = Be^{-\theta_D/bT} . T^3 \omega^2, \tag{5.21}$$

where B and b are constants, θ_D is the Debye temperature, T is the actual temperature, and ω is the frequency. In the case of phonon–impurity scattering, relaxation time can be calculated by

$$\tau_I^{-1} = A\omega^4, \tag{5.22}$$

where A is a constant.

The total relaxation time can be calculated by the harmonic mean of the umklapp scattering and phonon–impurity scattering.

$$\frac{1}{\tau} = \frac{1}{\tau_u} + \frac{1}{\tau_I}. \tag{5.23}$$

This is called the Harmonic rule or the Matthiessen rule.

The Boltzmann transport equation can solve all kinds of nanoscale transport problems. However, this equation has some limitations.

The first limitation is that the BTE is limited to a particle-based approach. For a particle approach to be valid, the wavelength should be much smaller than the characteristic dimension. Otherwise, the wave effects need to be considered. This means that the transport equation that can study the wave nature and the corresponding transport of waves has to be formulated. This approach is entirely different from heat transport.

The other significant limitation is that the Boltzmann transport equation can be used to solve for a dilute system of particles such as phonons, photons, electrons, and gas molecules. However, it cannot be used to solve for liquids because, in liquids, strong intermolecular forces exist, which are entirely different from the molecular forces in ideal gases. Therefore, if the BTE has to be used to solve for liquids, the collision term has to be modified to account for more complex intermolecular interactions. The modified equation is called the Enskog equation.

The BTE with the relaxation time approximation is found to be quite adequate and accurate for capturing physical phenomena. The classical continuum constitutive relations such as Fourier's equation, Fick's law, Ohm's law, and Newton's shear stress law can be derived by using the BTE. The BTE is an integral differential equation, that is, it can be written as the integral of a particular distribution function. Therefore, it is difficult to find out a simple analytical solution. This is why certain assumptions are used to simplify the analysis.

We know that

$$f - f_{eq} = Ce^{-\frac{t}{\tau}}.$$

For large values of $\frac{t}{\tau}$, the $f - f_{eq}$ value is minimal. Therefore, to derive the continuum equations, one should assume that the difference between the nonequilibrium and equilibrium distribution functions should be equal to another function g, which is small compared to the equilibrium distribution function, i.e.:

$$f - f_{eq} = g. \tag{5.24}$$

This is the first and foremost assumption that should be made before deriving the classical constitutive relations from the BTE. The other assumption is that transient terms can be neglected because the classical constitutive relations such as Fourier's law and Newton's shear stress law are based on steady-state processes.

Therefore, rewriting the BTE in terms of g and f_{eq}, we obtain

$$\frac{\partial g}{\partial t} + \frac{\partial f_{eq}}{\partial t} + \vec{v} \cdot \nabla_r g + \vec{v} \cdot \nabla_r f_{eq} + \frac{\vec{F}}{m} \cdot \nabla_v g + \frac{\vec{F}}{m} \cdot \nabla_v f_{eq} = \frac{-g}{\tau}. \tag{5.25}$$

Therefore, based on these assumptions, transient terms can be neglected. Also, g is small compared to f_{eq}. The change in distribution function in either the physical space or the momentum space should be quite small compared to the relative change in the equilibrium distribution function. Therefore, the gradients $\nabla_r g$ or $\nabla_v g$ should be smaller than $\nabla_r f_{eq}$ and $\nabla_v f_{eq}$. Under this assumption, the other terms involving the gradient of g can be neglected. Therefore, subjecting g, we obtain

$$g = -\tau \left(\vec{v} \cdot \nabla_r f_{eq} + \frac{\vec{F}}{m} \cdot \nabla_v f_{eq} \right). \tag{5.26}$$

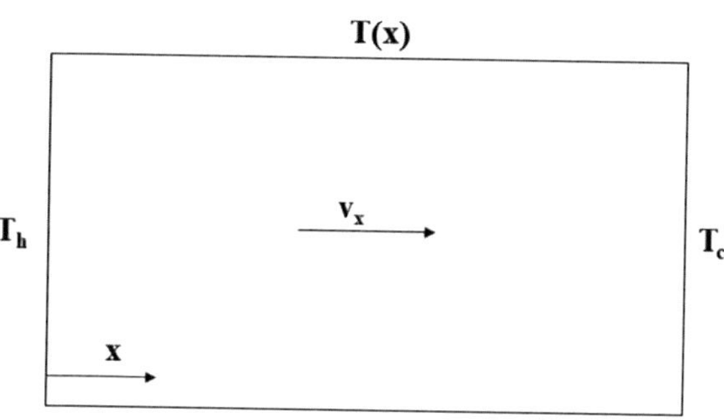

Fig. 5.5 Solid rectangular domain

5.1.5 Classical Constitutive Laws: Fourier's Law, Ohm's Law, and Newton's Shear Stress Law

5.1.5.1 Fourier's Law

Fourier's law of heat conduction can be derived from the Boltzmann transport equation. Consider a domain in which one end is maintained at a higher temperature (T_h) and the other end is maintained at a lower temperature (T_c), as shown in Fig. 5.5. There is a temperature variation along the x-direction. Assume that the domain is filled with phonon gas. The energy of each quantum state can be assumed to be $\hbar\omega$ and the nonequilibrium distribution function can be denoted by f. A specified number of phonons will occupy each quantum state. Therefore, the energy distribution in a quantum state is given by $\hbar\omega f$. In the wave vector space, the total energy can be obtained by summing up the energy distribution in a quantum state over all the quantum states:

$$E = \sum_{k_z}\sum_{k_y}\sum_{k_x} \hbar\omega f. \tag{5.27}$$

Assume that the phonons are moving with a velocity v_x. The net heat flux along the x-direction is given by

$$Q(x) = \frac{1}{V}\sum_{k_z}\sum_{k_y}\sum_{k_x} v_x \hbar\omega f. \tag{5.28}$$

The summation can be converted into an integral as

$$
\begin{aligned}
Q(x) &= \frac{1}{L^3}\int_{k_z}\int_{k_y}\int_{k_x} v_x \hbar\omega f \frac{dk_x dk_y dk_z}{\left(2\pi/L\right)^3} \\
&= \frac{1}{L^3}\int_{k_z}\int_{k_y}\int_{k_x} v_x \hbar\omega f \frac{k^2 dk \sin\theta\, d\theta\, d\phi}{\left(2\pi/L\right)^3}.
\end{aligned}
\tag{5.29}
$$

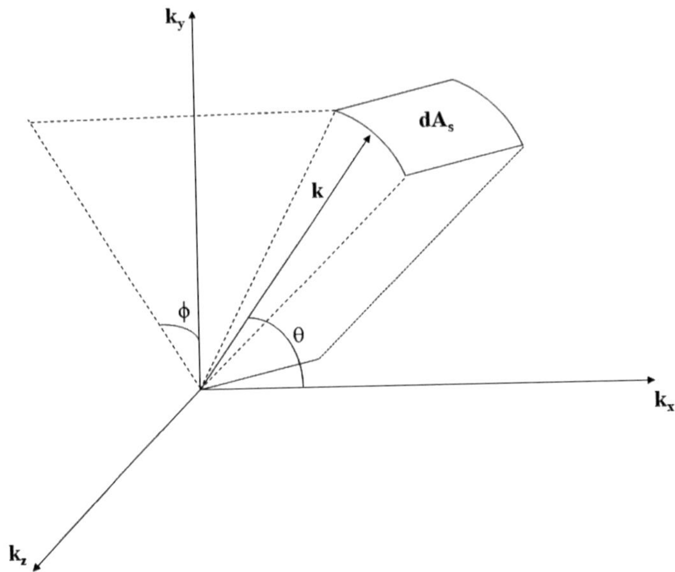

Fig. 5.6 Portion of a solid sphere in the $k_x k_y k_z$ space

Consider a solid sphere in the $k_x k_y k_z$ space. The volume of each of these unit cells in a 3-D wave vector space $k_x k_y k_z$ is $\left(\frac{2\pi}{L}\right)^3$. The number of quantum states within the sphere can be obtained by dividing the total volume of the sphere by the volume of a single unit cell. Consider a portion of the sphere having a wave vector k, as shown in Fig. 5.6. The angle made by this portion with the x-axis is the polar angle θ. Projection of the wave vector k onto the yz-plane results in an azimuthal angle ϕ with the y-axis. The surface area (dA_s) of the portion of the sphere is equal to $k^2 \sin\theta d\theta d\phi$, which gives the solid angle $d\theta$. The volume of the part of the sphere can be obtained by multiplying $k^2 d\theta$ with dk. To replace the discrete summation with an integral, divide $dk_x dk_y dk_z$ by $\left(\frac{2\pi}{L}\right)^3$, which gives the total number of quantum states. To convert the $k_x k_y k_z$ space into the spherical coordinate system, $dk_x dk_y dk_z$ can be written as $k^2 dk \sin\theta d\theta d\phi$. The total volume is equal to L^3. The limits of θ and ϕ vary from 0 to π and 0 to 2π, respectively.

Therefore, the resulting expression for heat flux ($Q(x)$) can be written as

$$Q(x) = \frac{1}{(2\pi)^3} \int_0^{2\pi} \int_0^{\pi} \int_0^{\omega_{max}} \hbar\omega v_x f \sin\theta d\theta d\phi (k^2 dk). \qquad (5.30)$$

The number of quantum states per unit frequency interval per unit volume is the density of states per unit volume. The density of states can be expressed as

$$D(\omega) = \frac{dN}{V d\omega} = \frac{4\pi k^2 dk}{L^3 \left(\frac{2\pi}{L}\right)^3 d\omega},$$ (5.31)

where $4\pi k^2$ is the surface area of the annular strip, the volume of each quantum state is $\left(\frac{2\pi}{L}\right)^3$, and $d\omega$ is the unit frequency interval. From the expression for the density of states,

$$k^2 dk = D(\omega) d\omega (2\pi^2).$$ (5.32)

Substituting $k^2 dk$ into the expression for heat flux, we obtain

$$Q(x) = \frac{1}{4\pi} \int_0^{2\pi} \int_0^{\pi} \int_0^{\omega_{max}} \hbar\omega v_x f \sin\theta \, d\theta \, d\phi \, D(\omega) d\omega.$$ (5.33)

The x-component of velocity $v_x = v\cos\theta$.

The nonequilibrium distribution function f can be written with respect to f_{eq} because it is much easier to integrate. f_{eq} does not change with respect to θ and ϕ, and it is only a function of temperature. On expressing the nonequilibrium distribution function f in terms of f_{eq}, the expression for heat flux becomes

$$Q(x) =$$

$$\frac{1}{4\pi} \int_0^{\omega_{max}} d\omega \left[\int_0^{2\pi} \left\{ \int_0^{\pi} v\cos\theta \hbar\omega \left[f_{eq} - \frac{d f_{eq}}{dT} \frac{\partial T}{\partial x} v\cos\theta \right] D(\omega)\sin\theta \, d\theta \right\} d\phi \right].$$ (5.34)

From the BTE,

$$\nabla_r \cdot f = \frac{d f_{eq}}{dx}$$

and

$$\frac{d f_{eq}}{dx} = \frac{d f_{eq}}{dT} \frac{\partial T}{\partial x}$$

because f_{eq} is a function of temperature. On further simplification, the expression for heat flux becomes

$$Q(x) = -\frac{1}{2} \frac{dT}{dx} \int_0^{\omega_{max}} d\omega \int_0^{\pi} \tau v^2 \cos^2\theta \sin\theta \hbar\omega \frac{d f_{eq}}{dT} d\theta.$$ (5.35)

Here,

$$\int_0^{\pi} \sin\theta \cos^2\theta \, d\theta = \frac{2}{3}$$

and

$$C_v = \int_0^{\omega_{max}} \frac{d}{dt}\left(\hbar\omega f_{eq} D(\omega)\right) d\omega,$$

where C_v is the volumetric heat capacity. This is called the microstate heat capacity. The macroscale heat capacity can be obtained by integrating the microscale heat capacity over all the microstates.

Therefore, the expression for heat flux becomes

$$Q(x) = -\frac{1}{3}\frac{dT}{dx}\int_0^{\omega_{max}} d\omega\left\{\tau v^2 C_v(\omega)\right\}. \qquad (5.36)$$

By comparing with $Q = -k\frac{dT}{dx}$, the expression for thermal conductivity (k) is given by

$$k = \frac{1}{3}\int_0^{\omega_{max}} \tau v^2 C_v(\omega) d\omega. \qquad (5.37)$$

Note that the relaxation time (τ) and velocity (v) are functions of frequency. Hence, they cannot be written outside of the integral.

If the relaxation time (τ) and velocity (v) are assumed to be constant and independent of ω, then the expression for thermal conductivity becomes

$$k = \frac{1}{3}\tau v^2 C_v. \qquad (5.38)$$

In terms of the mean free path,

$$k = \frac{1}{3}\Lambda v C_v. \qquad (5.39)$$

5.1.5.2 Ohm's Law

Analogously, Ohm's law can also be deduced. In this case, the equilibrium distribution function (f_{eq}) takes the form of the Maxwell–Boltzmann distribution, which is continuous. In this case, the expression for electron charge flux or current flux has to be obtained. The electric field is associated with the charge flux or current flux. Therefore, the force corresponding to the current flux is called the Lorentz force, and it is primarily due to the flow of electrons. Instead of $\hbar\omega$, the charge of an electron (e^-) has to be substituted. To derive Ohm's law, assume that the entire metal or semiconductor material is isothermal.

The expression for the current flux (J_x) can be written in integral form as

$$J_x = \frac{1}{V} \int\limits_{k_z} \int\limits_{k_y} \int\limits_{k_x} v_x(-e) f \frac{dk_x dk_y dk_z}{\left(2\pi/L\right)^3}. \tag{5.40}$$

The unit of current flux or charge flux is Amp/m^2. The expression for current flux can be converted from wave vector space into energy space through the density of states. Therefore, the expression becomes

$$J_x = \frac{1}{4\pi} \int\limits_0^{2\pi} \int\limits_0^{\pi} \int\limits_0^{\infty} v_x(-e) f D(E) \sin\theta \, d\theta \, d\phi \, dE. \tag{5.41}$$

In this expression, f can be expressed in terms of f_{eq}. Writing f in terms of f_{eq},

$$f = f_{eq} - \tau \left[v_x \frac{\partial f_{eq}}{\partial x} - \frac{eE_x}{m} \frac{\partial f_{eq}}{\partial v_x} \right], \tag{5.42}$$

where f_{eq} is a function of x only through the gradient of the Fermi energy level E_f. In the case of the Fermi–Dirac distribution function, f_{eq} is a function of energy and temperature. Therefore, $\frac{\partial f_{eq}}{\partial x}$ can be expressed as

$$\frac{\partial f_{eq}}{\partial x} = \frac{\partial f_{eq}}{\partial E_f} \frac{dE_f}{dx}, \tag{5.43}$$

where E_f is the Fermi energy level. Similarly, $\frac{\partial f_{eq}}{\partial v_x}$ can be expressed as

$$\frac{\partial f_{eq}}{\partial v_x} = \frac{\partial f_{eq}}{\partial E} \frac{dE}{dv_x}. \tag{5.44}$$

The second term on the right-hand side is the Lorentz force term. The charge of the electron (e^-) multiplied with the electric field (E) gives the Lorentz force (F_x).

Similarly, $\frac{\partial f_{eq}}{\partial v_x}$ is a function of v_x through E. From the dispersion relations, E can be written as

$$E = E_c + \frac{\hbar^2 k^2}{2m^*}, \tag{5.45}$$

where E_c is the energy at the beginning of the conduction band and m^* is the effective mass of the crystal. Substituting $\hbar k = mv$, the expression becomes

$$E = E_c + \frac{1}{2} m^* \left(v_x^2 + v_x^2 + v_x^2 \right). \tag{5.46}$$

Differentiating E with respect to v_x,

$$\frac{\partial E}{\partial v_x} = m^* v_x.$$ (5.47)

Substituting $\frac{\partial E}{\partial v_x}$, the expression for f becomes

$$f = f_{eq} - \tau \left[v_x \frac{\partial f_{eq}}{\partial E_f} \frac{dE_f}{dx} - eE_x \frac{\partial f_{eq}}{\partial E} \right].$$ (5.48)

Differentiating the Fermi–Dirac distribution function with respect to E and E_f,

$$\frac{\partial f_{eq}}{\partial E_f} = -\frac{\partial f_{eq}}{\partial E}.$$ (5.49)

Therefore,

$$f = f_{eq} + \tau v_x \left[\frac{dE_f}{dx} + eE_x \right] \frac{df_{eq}}{dE}.$$ (5.50)

Substituting f into the expression for current density (J_x), we obtain

$$J_x = \frac{-e}{4\pi} \int_0^{2\pi} \int_0^{\pi} \int_0^{\infty} v_x^2 \tau \left[\frac{dE_f}{dx} + eE_x \right] \frac{df_{eq}}{dE} D(E) dE \sin\theta d\theta d\phi.$$ (5.51)

Here, $v_x^2 = v^2 \cos^2\theta$ and $\int_0^{\pi} \cos\theta \sin\theta d\theta = \frac{2}{3}$.

Writing the term $\frac{dE_f}{dx} + eE_x$ outside the integral,

$$J_x = -\frac{e}{3} \left[\frac{dE_f}{dx} + eE_x \right] \int_0^{\infty} v^2 \tau \frac{df_{eq}}{dE} D(E) dE.$$ (5.52)

This expression can be applied to both metals and semiconductors. In the case of metals, $\frac{dE_f}{dx} = 0$ because of the large values of electron number density.

Therefore, the current density of metals is given by

$$J_e = -\frac{e^2}{3} E_x \int_0^{\infty} v^2 \tau \frac{df_{eq}}{dE} D(E) dE.$$ (5.53)

According to Ohm's law, current flux is equal to electrical conductivity times the electric field, i.e.:

$$J_e = \sigma E_x.$$ (5.54)

Therefore,

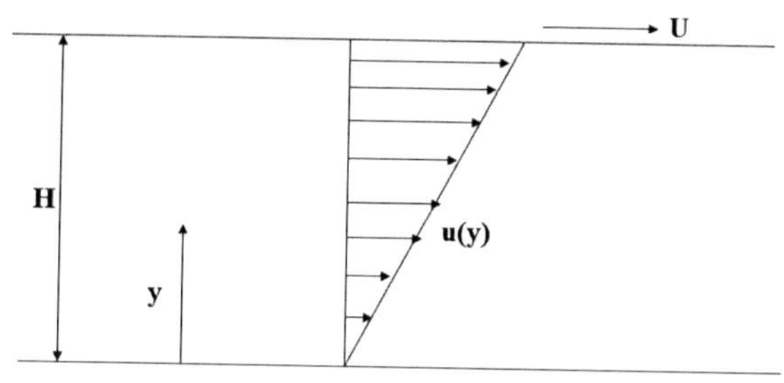

Fig. 5.7 Couette flow between two parallel plates

$$\sigma = -\frac{e^2}{3} \int_0^\infty v^2 \tau \frac{d f_{eq}}{dE} D(E) dE. \tag{5.55}$$

This is the expression for electrical conductivity.

Thermoelectric effects can be explained with a simple example. Consider a non-isothermal surface in which the simultaneous transport of charge flux and heat flux takes place. Since the surface is nonisothermal, a temperature gradient (dT/dx) exists along with the electron flux (dE_f/dx).

Therefore,

$$\frac{d f_{eq}}{dx} = \frac{\partial f_{eq}}{\partial E_f}\frac{dE_f}{dx} + \frac{\partial f_{eq}}{\partial T}\frac{dT}{dx}. \tag{5.56}$$

Here, the first and second terms on the right-hand side are related to the charge transport and heat transport, respectively. Substituting this into the expression for current flux gives the nonequilibrium distribution function. A similar procedure can be adopted to calculate the Seebeck and Peltier coefficients.

5.1.5.3 Newton's Shear Stress Law

Newton's shear stress law can be derived in an analogous way for gas molecules. According to Newton's shear stress law,

$$\tau_{xy} = \mu \frac{\partial u}{\partial y}. \tag{5.57}$$

Consider the Couette flow between two parallel plates, as shown in Fig. 5.7. The top plate is moving with a velocity U and the bottom plate is fixed. The separation distance between the plates is H. The variation of velocity between the top and bottom plates is linear. The molecular velocities in the x- and y-directions are v_x and v_y, respectively.

The Maxwell–Boltzmann distribution function is given by

$$f_{eq} = n\left(\frac{m}{2\pi k_B T}\right)^{3/2} e^{\left(\frac{-m((v_x - u(y))^2 + v_y^2 + v_z^2)}{2k_B T}\right)}, \tag{5.58}$$

where n is the number density of gas molecules and m is the mass of the molecules. In this case, the bulk velocity of molecules in the x-direction is u, which is a function of y. For the Couette flow,

$$u(y) = \frac{Uy}{H}. \tag{5.59}$$

The classical Maxwell–Boltzmann distribution does not account for the bulk motion. Therefore, the velocity should be corrected in the x-direction by subtracting the bulk velocity from the molecular velocity. V_x is the molecular velocity and $u(y)$ is the bulk velocity. Only molecular velocities exist in the other directions. Therefore, in the case of flow transport, the Maxwell–Boltzmann distribution becomes the displaced Maxwell–Boltzmann distribution.

The relation between f and f_{eq} can be written as

$$f = f_{eq} - \tau \left(\vec{v}\, \nabla_r f_{eq}\right). \tag{5.60}$$

On expanding, we obtain

$$f = f_{eq} - \tau \left(v_x \frac{\partial f_{eq}}{\partial x} + v_y \frac{\partial f_{eq}}{\partial y}\right). \tag{5.61}$$

Here, $\frac{d f_{eq}}{dy}$ can be expressed as

$$\frac{\partial f_{eq}}{\partial y} = \frac{\partial f_{eq}}{\partial u} \frac{du}{dy}. \tag{5.62}$$

This nonequilibrium distribution function f should be substituted into the expression for shear stress.

The shear stress can be obtained by calculating the momentum flux. In this case, the x-momentum (mv_x) is considered in the y-direction. Therefore, the x-momentum is multiplied with v_y and the corresponding nonequilibrium distribution function (f). Then, the expression must be integrated over the entire momentum space $v_x v_y v_z$ to obtain shear stress (τ):

$$\tau_{xy} = \int_{-\infty}^{\infty}\int_{-\infty}^{\infty}\int_{-\infty}^{\infty} v_y m v_x f \, dv_x dv_y dv_z. \tag{5.63}$$

The limits of integration for v_x, v_y, and v_z vary from $-\infty$ to ∞.

According to Newton's law of viscosity, shear stress is expressed by

$$\tau_{xy} = \mu \frac{\partial u}{\partial y}.$$ (5.64)

The expression for dynamic viscosity can be obtained by substituting the nonequilibrium distribution function f into the shear stress expression and then comparing with Newton's law of viscosity.

Finally, dynamic viscosity is expressed by

$$\mu = \frac{1}{4d^2}\sqrt{\frac{mk_B T}{\pi}}.$$ (5.65)

Thus, all the constitutive continuum relations can be derived analogously by using the Boltzmann transport equation.

5.1.6 Assumptions Made in the Derivation of Constitutive Continuum Relations

However, there are some limitations of the classical constitutive relations. These continuum equations are not valid at high Knudsen numbers. Some assumptions are made in deriving all these constitutive continuum relations. The assumptions are as follows:

1. The deviation (g) between the nonequilibrium (f) and equilibrium (f_{eq}) distribution functions is substantially smaller than the equilibrium distribution function (f_{eq}). This is the basis of deriving all continuum equations. Therefore, as the deviation increases, the local nonequilibrium conditions increase. Therefore, the function g indicates the deviation of f from f_{eq}:

$$g << f_{eq}.$$ (5.66)

2. The gradients of the deviation function (g) are much smaller than those of the equilibrium distribution function:

$$\nabla_r g, \nabla_v g << \nabla_r f_{eq}, \nabla_v f_{eq}.$$ (5.67)

3. The physical time (t) scales are much higher than the relaxation time (τ) scales. This means that

$$\frac{\partial g}{\partial t} << \frac{g}{\tau}.$$ (5.68)

The third condition may not be satisfied, even if the Knudsen number range is in the continuum regime. This can be explained with a simple example. Consider a laser pulse irradiated to a metal or semiconductor surface. The pulse width of the laser is of the order of picoseconds (10–12 sec) or femtoseconds (10–15 sec). When the pulse strikes the surface, the mean free path is in the range from 20 nm to a few hundreds of nanometers. In general, all the relaxation timescales are of the order of nanometers. In this case, the third condition is clearly violated, but the Knudsen number perfectly lies in the continuum regime.

In the case of transient heat conduction, the semi-infinite approximation is used. Consider an example of temperature variation from the core of the earth to the upper layer. Assume that the upper layer is maintained at a temperature of T_{wall}. The temperature at the core, which is several thousands of kilometers from the upper surface, is entirely different from the upper surface temperature (T_{wall}) of the earth because there is no source of energy at the core. This is why the core is unable to sense the temperature of the upper layer within a few seconds or minutes. The core temperature always maintains its initial temperature. In this case, it is not necessary to solve the transient conduction rigorously through the entire body. The initial condition is imposed on the core as a boundary condition. Therefore, the core is always at the initial temperature.

The heat equation is given by

$$\frac{\partial T}{\partial t} = \alpha \frac{\partial^2 T}{\partial x^2}. \tag{5.69}$$

The boundary conditions are as follows: at $x = 0$, $T = T_{wall}$ and at $x = \infty$, $T = T_i$ (initial temperature).

The solution for this transient heat conduction is given by

$$T(x, t) = e^{\frac{-x}{\sqrt{\alpha t}}}. \tag{5.70}$$

This signifies that temperature decays exponentially with x. At large values of x, the temperature will have a finite value. This means that the speed of propagation of heat is not equal to the speed of sound in the medium. On applying the temperature or heat flux at one end, the other end can sense and respond to the corresponding temperature or heat flux. It is practically not feasible because the energy carriers, in this case, would be phonons or electrons, which can only travel with a finite speed.

The solution to the transient heat conduction problem is obtained based on Fourier's equation. Therefore, using Fourier's equation, it is obvious that the speed of propagation of heat is infinite and the other end is still able to sense the finite value of temperature. Hence, it is a clear illustration of the fallacy of using Fourier's constitutive relationship. To overcome this fallacy, Cattaneo's equation is used, which is a modification of Fourier's equation.

In general, Fourier's equation says that the heat flux is equal to $-k\frac{\partial T}{\partial x}$. On adding the correction term $\tau\frac{\partial Q}{\partial t}$ on the left-hand side of Fourier's equation, Cattaneo's equation is obtained as follows:

$$\tau \frac{\partial Q}{\partial t} + Q = -k \frac{\partial T}{\partial x}.$$

(5.71)

In this equation, the additional term is the relaxation time of the energy carriers. The continuum equations do not describe the collision of energy carriers. The collision term accounts for the finite speed of propagation of heat. Therefore, an additional term involving the relaxation time (τ) is added to Fourier's equation. The additional term will vanish in the limiting case where the relaxation time (τ) approaches 0. Therefore, in the limiting case, Cattaneo's equation becomes Fourier's equation. The hyperbolic heat conduction equation can be obtained by substituting Cattaneo's equation into the heat conduction equation. The equation for heat conduction is given by

$$\rho C \frac{\partial T}{\partial t} = -\nabla \vec{Q}.$$

(5.72)

All conservation equations such as conservation of mass, conservation of momentum, and conservation of energy can be derived from the Boltzmann transport equation by multiplying the quantities such as mass, momentum, and energy with the BTE and integrating over the entire momentum space:

$$\int X \left(\frac{\partial f}{\partial t} + v \nabla_r f + \frac{F}{m} \nabla_v f \right) d^3 V.$$

(5.73)

Here, X can be replaced with mass, momentum, and energy to obtain the respective conservative equation.

5.2 Size Effects

Size effects are predominant in the case of nanoscale heat transport. They have a significant impact on electrical conductivity and thermal conductivity. In nanoscale transport processes, the Boltzmann transport equation is the fundamental equation from which one can derive all the classical constitutive relations.

The BTE can be expressed as

$$\frac{\partial g}{\partial t} + \frac{\partial f_{eq}}{\partial t} + \vec{v} \cdot \nabla_r g + \vec{v} \cdot \nabla_r f_{eq} + \frac{\vec{F}}{m} \cdot \nabla_v g + \frac{\vec{F}}{m} \cdot \nabla_v f_{eq} = \frac{-g}{\tau}.$$

(5.74)

Several assumptions were made when deriving the continuum equations, one of which is that the physical timescales are much higher than the relaxation timescales. Therefore, the scattering term is greater than $\frac{\partial g}{\partial t}, \frac{\partial f_{eq}}{\partial t}$, as shown in the following equation:

$$\frac{\partial g}{\partial t}, \frac{\partial f_{eq}}{\partial t} << \frac{g}{\tau}. \tag{5.75}$$

We also invoked the assumption that the perturbation function g should be small compared to f_{eq} and the respective gradients of g should be smaller than $\nabla_r f_{eq}$ and $\nabla_v f_{eq}$.

Therefore, f can be expressed in terms of f_{eq} as

$$f = f_{eq} - \tau \left(\vec{v} \cdot \nabla_r f_{eq} + \frac{\vec{F}}{m} \cdot \nabla_v f_{eq} \right). \tag{5.76}$$

In the present case, our primary intention is to study the transport of nanostructures. Therefore, all these assumptions cannot be made in nanoscale energy transport. However, one can still assume with respect to timescales because, although the Knudsen numbers are high, the principal focus is on solving the problems where the physical timescales are higher than the relaxation timescales.

Size effects arise due to the reduction in the size of the structure and not due to the disparity in the timescales. Therefore, $\frac{\partial g}{\partial t}$ and $\frac{\partial f_{eq}}{\partial t}$ are much smaller than the collision term on the right-hand side of the BTE:

$$\frac{\partial g}{\partial t}, \frac{\partial f_{eq}}{\partial t} << \frac{g}{\tau}. \tag{5.77}$$

This is one of the crucial assumptions. The other critical assumption is that the gradient of the perturbation function in the momentum space is also smaller than the corresponding change with respect to f_{eq}. Therefore,

$$\frac{\vec{F}}{m} \cdot \nabla_v g << \frac{\vec{F}}{m} \cdot \nabla_v f_{eq}. \tag{5.78}$$

After invoking these two assumptions, the BTE can be rewritten in 2-D Cartesian coordinates as

$$v_x \frac{\partial f_{eq}}{\partial x} + v_y \frac{\partial f_{eq}}{\partial y} + v_x \frac{\partial g}{\partial x} + v_y \frac{\partial g}{\partial y} + \frac{F_x}{m}\frac{\partial f_{eq}}{\partial v_x} + \frac{F_y}{m}\frac{\partial f_{eq}}{\partial v_y} = \frac{-g}{\tau}. \tag{5.79}$$

Here, the unsteady and advection terms in both physical and momentum space are neglected. Let us now consider two kinds of transport processes.

One is transport parallel to thin films and the other is transport perpendicular to thin films. Thin films are nanostructures. Consider a thin film having a thickness of "d" of the order of several nanometers, as shown in Fig. 5.8. The coordinate system is fixed at the bottom plane such that the velocity vectors in the x- and y-directions

Fig. 5.8 Transport parallel
to thin films

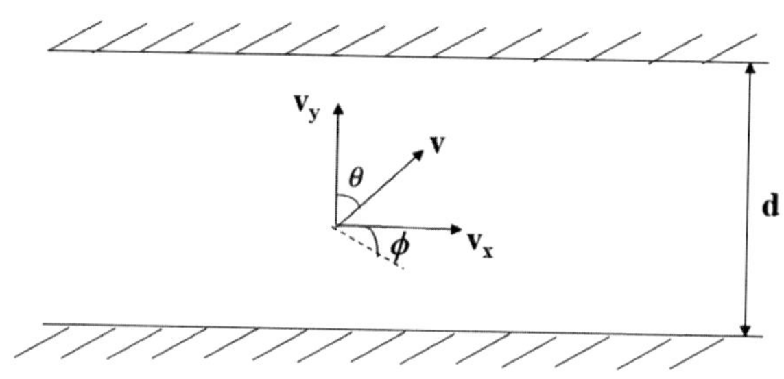

are v_x and v_y, respectively. The resultant of v_x and v_y is v, which makes an angle θ with the vertical direction. θ is called the polar angle. The projection of v on the xz-plane makes an angle of ϕ with the horizontal direction. ϕ is called the azimuthal angle.

5.2.1   Transport Parallel to Thin Films

In the case of transport parallel to thin films, the temperature gradient or the gradient of Fermi energy level exists in the horizontal direction such that the electrons move along the x-direction.

Therefore, $\frac{\partial f_{eq}}{\partial y}$ is smaller than the corresponding gradient of the deviation function in the same direction:

$$\frac{\partial f_{eq}}{\partial y} << \frac{\partial g}{\partial y}. \tag{5.80}$$

Since the transport is in the x-direction, $\frac{\partial f_{eq}}{\partial x}$ is far higher than the gradient of the perturbation function, i.e.:

$$\frac{\partial f_{eq}}{\partial x} >> \frac{\partial g}{\partial x}. \tag{5.81}$$

Similarly, $\frac{\partial f_{eq}}{\partial v_x}$ is far higher than the corresponding gradient of the deviation function:

$$\frac{\partial f_{eq}}{\partial v_x} >> \frac{\partial f_{eq}}{\partial v_y}. \tag{5.82}$$

Depending on the direction of the transport, one can say that the order of magnitude of the derivative of f_{eq} is significant in the x-direction and non-significant in the y-direction.

After invoking all these assumptions, it can be seen that the BTE reduces to

$$v_x \frac{\partial f_{eq}}{\partial x} + v_y \frac{\partial g}{\partial y} + \frac{F_x}{m} \frac{\partial f_{eq}}{\partial v_x} = \frac{-g}{\tau}. \tag{5.83}$$

5.2.2   *Transport Perpendicular to Thin Films*

On maintaining the temperature gradient or an electric field perpendicular to the thin film, the gradient of the equilibrium distribution function in the y-direction is higher than the corresponding gradient of the perturbation function in the same direction. Therefore,

$$\frac{\partial f_{eq}}{\partial y} >> \frac{\partial g}{\partial y}. \tag{5.84}$$

Similarly, $\frac{\partial f_{eq}}{\partial x}$ is much smaller than $\frac{\partial g}{\partial x}$, i.e.:

$$\frac{\partial f_{eq}}{\partial x} << \frac{\partial g}{\partial x}. \tag{5.85}$$

$\frac{\partial f_{eq}}{\partial v_y}$ is larger than the derivative of perturbation function, i.e.:

$$\frac{\partial f_{eq}}{\partial v_y} >> \frac{\partial f_{eq}}{\partial v_x}. \tag{5.86}$$

Therefore, by incorporating these assumptions, the BTE becomes

$$v_y \frac{\partial f_{eq}}{\partial y} + v_x \frac{\partial g}{\partial x} + \frac{F_y}{m}\frac{\partial f_{eq}}{\partial v_y} = \frac{-g}{\tau}. \tag{5.87}$$

In the case of transport parallel to thin films, even though the transport is in the x-direction, size effects exist in the y-direction because of the confinement. This is because, in the nanoscale, electron–boundary scattering takes place along with electron–electron scattering, which brings about quite a substantial distortion in the nonequilibrium function. Therefore, even when transport is along the plane, distortion can happen in the perpendicular direction because of the scattering mechanism. Hence, the term $\frac{dg}{dy}$ is still strong in this case, which is primarily responsible for the size effect.

In the case of transport perpendicular to thin films, it is obvious that the temperature gradient exists in the y-direction because transport is in the same direction. Therefore, f_{eq} varies predominantly along that direction. There is no variation in the distortion function along the x-direction. In this case, both the distortion function and f_{eq} vary along the y-direction. In the case of transport parallel to thin films, f_{eq} varies along the x-direction, but the distortion function varies along the y-direction, which causes the size effect.

Consider phonon or electron transport parallel to thin films. Writing like terms on both sides, the BTE becomes

$$\tau v \cos\theta \frac{dg}{dy} + g = -\tau \frac{F_x}{m} \frac{df_x}{dv_x} - \tau v_x \frac{df_{eq}}{dx}$$

$$= -S_0(x).$$

(5.88)

The terms on the right-hand side can be denoted by $S_0(x)$. Since we are interested in finding the solution of g as a function of y, which contains the information on the size effects, the terms involving y are written on the left side and the terms involving x on the right side.

The complete solution to an ordinary differential equation (ODE) can be obtained by adding the complementary function to the particular integral. The complementary function is the solution to the homogeneous differential equation.

The complementary function can be obtained by solving

$$\tau v \cos\theta \frac{dg}{dy} + g = 0.$$

(5.89)

Therefore, the complementary function can be written as

$$g(y) = C e^{-\frac{y}{\tau v \cos\theta}}.$$

(5.90)

It is known that the confinement is along the y-direction with a few nanometers. Since g is a function of y, the distortion of g takes place along the y-direction, which brings about the size effect in electrical conductivity or thermal conductivity. Therefore, the complete solution is given by

$$g(y) + S_0 = C e^{-\frac{y}{\tau v \cos\theta}}.$$

(5.91)

The particular integral is given by

$$S_0(x) = -e\tau v_x \frac{df_{eq}}{dE}\left(E_x + \frac{1}{e}\frac{dE_f}{dx}\right).$$

(5.92)

If there is no confinement, the electron scattering at the boundaries is negligible. Therefore, electron–electron scattering is more predominant than electron–boundary scattering. The motion of an electron will not be affected due to confinement because the solution of BTE implies that the electron can travel in a spherical coordinate system encompassing all directions.

Therefore, electron—boundary scattering is the phenomenon that causes the gradient of the perturbation function along the y-direction, as shown in Fig. 5.9. With respect to the coordinate system, the velocity vectors along the x- and y-directions are v_x and v_y, respectively. v is the resultant velocity vector, which makes an angle θ with respect to the vertical direction. Therefore, if the scattering is in the upward direction, then θ varies from 0 to $\frac{\pi}{2}$, and for downward scattering, θ varies from $\frac{\pi}{2}$ to π, which is due to rotation of ϕ. This represents the occurrence of the scattering phenomenon in all the directions.

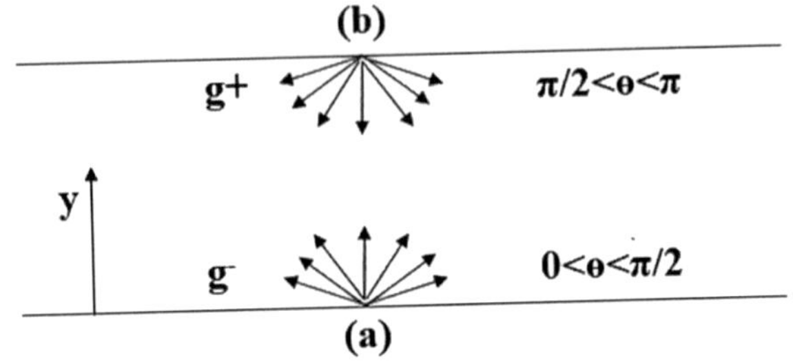

Fig. 5.9 Electron boundary scattering: **a** upward scattering; **b** downward scattering

Therefore, g is not only a function of y but also a function of θ. It involves both directional and physical coordinates. The scattering of electrons inside the thin film takes place in both the positive and negative y-directions. The electrons that reach the top boundary travel in the negative y-direction. Therefore, the direction cosines are negative since θ varies from $\frac{\pi}{2}$ to π. Similarly, the phonons that reach the bottom boundary travel in the positive y-direction. In this case, θ varies from 0 to $\frac{\pi}{2}$. Therefore, the direction cosines are positive. Hence, at every point, the perturbation function g can be divided into a positive g and a negative g. The azimuthal angle ϕ varies from 0 to 2π, which makes the distribution symmetric.

Therefore, the solution can be written in terms of g^+ and g^- and both are functions of y and μ. g^+ and g^- are the perturbation functions at the bottom and top boundaries, respectively. μ is the direction cosine, which is equal to $\cos\theta$. Therefore, g^+ can be written as

$$g^+(y, \mu) = C_1 e^{-\frac{y}{\tau v \mu}} - S_0. \tag{5.93}$$

Similarly, g^- can be written as

$$g^-(y, \mu) = C_2 e^{-\frac{y}{\tau v \mu}} - S_0. \tag{5.94}$$

Note that the constants C_1 and C_2 are not equal. In the case of rough boundaries, any incoming electron or phonon that strikes the boundary uniformly scatters with the same magnitude in different directions. This is called diffuse scattering. In the case of smooth boundaries, the electron or phonon makes an angle and gets reflected in such a way that the incident angle (θ_i) is equal to the reflected angle (θ_r). This is called specular scattering. Diffuse and specular scatterings are shown in Fig. 5.10.

Whenever the phonon distribution function encounters the diffuse interface, the nonequilibrium distribution function f becomes f_{eq} because the scattering is uniform in all directions, which is not true in the case of specular scattering. On applying the boundary condition at the bottom surface (at $y = 0$), f is equal to f_{eq} in diffuse scattering. Therefore, the perturbation function g is equal to 0. Similarly, at the top surface (at $y = d$) also, the nonequilibrium distribution function (f) is equal to the equilibrium distribution function (f_{eq}).

The constants C_1 and C_2 can be evaluated by substituting the boundary conditions at $y = 0, d$ into the expressions for g^+ and g^-. Therefore, $C_1 = S_0$ and $C_2 = S_0 e^{\frac{d}{\tau v \mu}}$.

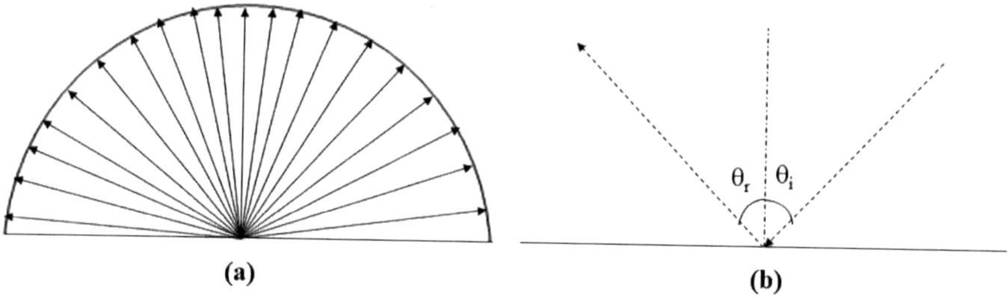

Fig. 5.10 **a** Diffuse scattering; **b** specular scattering

Hence, the expressions for g^+ and g^- become

$$g^+ = S_0 e^{-\frac{y}{\tau v \mu}} - S_0,\tag{5.95}$$

$$g^- = S_0 e^{\frac{d-y}{\tau v \mu}} - S_0.\tag{5.96}$$

The charge flux or current flux (J) for a single quantum state is given by

$$J = \frac{v_x(-e)f}{\left(2\pi/L\right)^3}.\tag{5.97}$$

Integrating over all the quantum states k_x, k_y, and k_z, we obtain

$$J_x = \frac{1}{V} \int\limits_{k_z} \int\limits_{k_y} \int\limits_{k_x} v_x(-e)f \frac{dk_x dk_y dk_z}{\left(2\pi/L\right)^3}.\tag{5.98}$$

The summation can be converted into an integral by using the density of states. Therefore, the momentum space can be converted into coordinates involving θ and ϕ, which can be written as

$$J_x = \int\limits_0^\infty dE \int\limits_0^{2\pi} d\phi \int\limits_0^\pi v_x(-e)f \frac{D(E)}{4\pi} \sin\theta d\theta.\tag{5.99}$$

Here, f can be expressed as

$$f = g + f_{eq}.\tag{5.100}$$

It is known that $v_x = v\cos\theta$, $\mu = \cos\theta$, and $d\mu = \sin\theta d\theta$.

Therefore, J becomes

$$J = -\frac{1}{4\pi} \int\limits_E eD(E)dE \int\limits_0^{2\pi} d\phi \left[\int\limits_{-1}^0 vg^-(y, -\mu) + \int\limits_0^1 vg^+(y, \mu) \right] d\mu. \quad (5.101)$$

Note that the new limits of integration become -1 to 1 and g can be expressed in terms of g^+ and g^-. The value of the integral that involves f_{eq} becomes 0 because f_{eq} is uniform in the θ- and ϕ-directions. Therefore, $\int f_{eq} \cos\theta \sin\theta d\theta d\phi$ is equal to 0.

Substituting g^+ and g^- in terms of S_0, the expression for J becomes

$$J = \frac{1}{4\pi} \int\limits_E eD(E)dE \int\limits_0^{2\pi} d\phi \left[v_x e\tau v_x \frac{d f_0}{dE} \left(E_x + \frac{1}{e}\frac{dE_f}{dx} \right) \right] \left[e^{-\frac{y}{v\tau\mu}} - 1 \right] d\mu$$

$$+ \int\limits_0^1 v_x e\tau v_x \frac{d f_0}{dE} \left(E_x + \frac{1}{e}\frac{dE_f}{dx} \right) \left[e^{\frac{d-y}{v\tau\mu}} - 1 \right] d\mu,$$

$$(5.102)$$

where S_0 can be expressed as

$$S_0 = -e\tau v_x \frac{\partial f_0}{\partial E} \left(E_x + \frac{1}{e}\frac{dE_f}{dx} \right). \quad (5.103)$$

The term v_x can be expressed as

$$v_x = v \sin\theta \cos\phi. \quad (5.104)$$

Therefore, the expression for J becomes

$$\frac{J}{E_x + \frac{1}{e}\frac{dE_f}{dx}} = \frac{1}{4} \int\limits_E e^2 v^2 \tau \frac{d f_{eq}}{dE} D(E)$$

$$\times \left[\int\limits_0^1 \left(1 - \mu^2\right) (e^{\frac{-y}{v\tau\mu}} - 1)d\mu + \int\limits_0^1 \left(1 - \mu^2\right) (e^{\frac{d-y}{v\tau\mu}} - 1)d\mu \right]. \quad (5.105)$$

The current flux divided by the electric field gives electrical conductivity $(-\sigma)$. Therefore, according to Ohm's law, electrical conductivity $(-\sigma)$ is given by

$$\frac{J}{E_x + \frac{1}{e}\frac{dE_f}{dx}}. \quad (5.106)$$

The expression for the bulk value of electrical conductivity is given by

Fig. 5.11 $\frac{\sigma}{\sigma_b}$ versus Kn

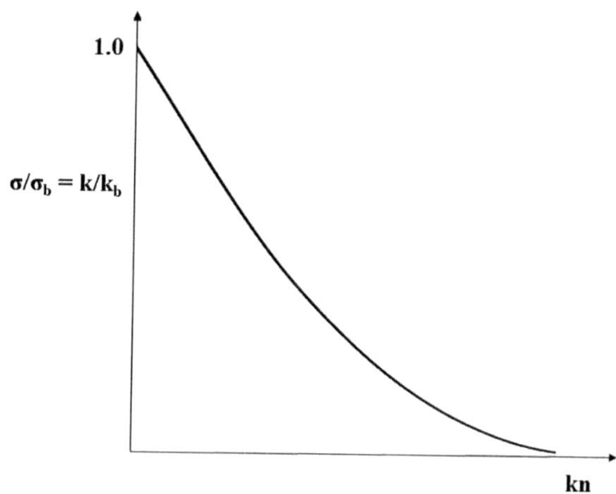

$$\sigma_b = -\frac{e^2}{3} \int \tau v^2 \frac{d f_{eq}}{d E} D(E) d E. \tag{5.107}$$

Therefore,

$$J_e = \sigma_b E_x. \tag{5.108}$$

Therefore, the ratio of σ to σ_b can be expressed as

$$\frac{\sigma}{\sigma_b} = 1 - \frac{3}{8\xi}(1 - 4E_3(\xi) - 4E_5(\xi)), \tag{5.109}$$

where $\xi = \frac{d}{\Lambda} = \frac{1}{kn}$. The integrals E_3 and E_5 can be evaluated separately by the following expression:

$$E_n(x) = \int_0^1 \mu^{n-2} e^{\frac{-x}{\mu}} d\mu. \tag{5.110}$$

Therefore, the resulting expression for the ratio of σ to σ_b gives the size effect in electrical conductivity for nanoscale transport. This is the simplest possible analytical solution for describing the size effects.

The size effect can be illustrated by plotting $\frac{\sigma}{\sigma_b}$ versus Kn, as shown in Fig. 5.11. If the Knudsen number is very small, the ξ term becomes large. Therefore, $\frac{\sigma}{\sigma_b}$ tends to 1. When the Knudsen number becomes large, then $\frac{\sigma}{\sigma_b}$ tends to 0. This is called the size effect.

Therefore, in a nanoscale transport phenomenon, there is a reduction in the electrical conductivity in the cases where the transport of electrons is parallel and perpendicular to the thin films because of boundary scattering. This scattering reduces the momentum of the electrons, which in turn reduces the amount of electron conduction.

Similarly, on replacing electrons with phonons, instead of an electric field, a temperature gradient develops in the x-direction. In this case, the ratio of $\frac{k}{k_b}$ gives the size effect in thermal conductivity.

Using the specular boundary condition, the ratio of $\frac{\sigma}{\sigma_b}$ is always equal to 1 for transport parallel to the thin film. In other words, the specular boundary condition does not affect the momentum of the electrons or phonons. Therefore, even if confinement is present, the specular boundary condition does not affect conductivity.

In the case of specular scattering at the boundaries, the general solution in terms of C_1 and C_2 remains the same. The constants C_1 and C_2 can be determined by applying the specular boundary condition. Therefore, g^+ and g^- are given as follows:

For $\mu > 0$,

$$g^+(y, \mu) = C_1 e^{-\frac{y}{\tau v \mu}} - S_0. \tag{5.111}$$

For $\mu < 0$,

$$g^-(y, \mu) = C_2 e^{-\frac{y}{\tau v \mu}} - S_0. \tag{5.112}$$

In the case of a specular surface, an electron or phonon is incident on the bottom boundary with an angle θ and scatters with the same angle (θ) in the other direction. Similarly, on the top boundary also, incident and reflected angles are the same. Therefore, at $y = 0$,

$$f^+(y = 0, \mu) = f^-(y = 0, -\mu), \tag{5.113}$$

where f^+ is the distribution function going out of the boundary in the positive y-direction and f^- is the distribution function coming into the boundary in the negative y-direction at $y = 0$.

Therefore, the incoming distribution function (f^-) is just reflected without altering the value of the distribution, but only the direction is changed. Similarly, at $y = d$,

$$f^-(y = d, -\mu) = f^+(y = d, \mu). \tag{5.114}$$

Therefore, f_{eq} is independent of the direction. Similarly, at $y = 0$,

$$g^+(y = 0) = g^-(y = 0), \tag{5.115}$$

and at $y = d$

$$g^-(y = d) = g^+(y = d). \tag{5.116}$$

Therefore, these two conditions are used to determine the constants C_1 and C_2. Hence, from the first condition at $y = 0$, we obtain $C_1 = C_2$.

The constant C_1 can be calculated by applying the second condition at $y = d$. Therefore,

$$C_1 e^{-\frac{d}{\tau v |\mu|}} - S_0 = C_1 e^{\frac{d}{\tau v |\mu|}} - S_0. \tag{5.117}$$

This equation is valid only when $C_1 = 0$. Therefore, $C_1 = C_2 = 0$. Substituting $C_1 = C_2 = 0$ into the expressions for g^+ and g^-, then

$$g^+ = g^- = -S_0(x). \tag{5.118}$$

Substituting g^+ and g^- into the expression for charge flux,

$$\frac{\sigma}{\sigma_b} = 1. \tag{5.119}$$

This expression shows that the value of electrical conductivity is the same as that of the bulk electrical conductivity. Therefore, momentum conservation is entirely satisfied in the case of specular scattering. In diffused scattering, there is an associated reduction in the energy of the emitted phonon or electron due to the distribution over the entire azimuthal or polar angles. This results in a consequent reduction in the conductivities. Therefore, it is essential to understand that electron transport parallel to the thin films need not always result in a reduction of conductivities. It depends on the type of boundary scattering at the boundaries.

Most of the real-time surfaces are neither entirely diffused nor entirely specular. Therefore, diffuse scattering and specular scattering can be categorized based on a parameter called specularity parameter (p). The p value for most of the real surfaces varies between 0 and 1.

The general expression for $\frac{\sigma}{\sigma_b}$ in terms of specularity parameter (p) is given by

$$\frac{\sigma}{\sigma_b} = 1 - \frac{3(1-p)}{2\xi} \int_0^1 \left[\frac{(\mu - \mu^3)(1 - e^{-\frac{\xi}{\mu}})}{(1 - pe^{-\frac{\xi}{\mu}})} \right] d\mu. \tag{5.120}$$

If $p = 1$, then it is completely specular scattering, and if $p = 0$, then it is completely diffused scattering. In the case of transport perpendicular to the thin film, the BTE becomes

$$v_y \frac{\partial f_{eq}}{\partial y} + v_y \frac{\partial g}{\partial y} + \frac{F_y}{m} \frac{\partial f_{eq}}{\partial v_y} = -\frac{g}{\tau}. \tag{5.121}$$

The distribution function can be obtained by solving this BTE.

The final expression for the distribution function in terms of intensity (I) is given by

$$2I_{eq} = I^+(0, \mu)E_2(\eta) + I^-(d, -\mu)E_2(\xi - \eta) + \int_0^\xi I_{eq}E_1(\eta' - \eta)d\eta', \tag{5.122}$$

where I^+ is the intensity of the phonon at $y = 0$ in the upward direction, I^- is the intensity of the phonon at $y = d$ in the downward direction, and η' is a dummy variable.

Here,

$$I_{eq} = \frac{\sigma T^4}{\pi}. \tag{5.123}$$

The integral functions can be expressed as

$$E_n(x) = \int_0^1 \mu^{n-2} e^{-x/\mu} d\mu. \tag{5.124}$$

These are popularly known as Fredholm integral functions. Therefore, the temperature distribution as a function of η has to be calculated. $\eta = \frac{y}{\Lambda}$ and $\xi = \frac{1}{Kn}$. On solving this expression in MATLAB or Excel sheets, the temperature can be obtained as a function of y.

The non-dimensional intensity I_{eq}^* can be expressed as

$$I_{eq}^* = \frac{I_{eq}(\eta) - I^-(d, -\mu)}{I^+(0, \mu) - I^-(d, -\mu)}. \tag{5.125}$$

I_{eq}^* can be obtained by non-dimensionalizing the local temperature with the boundary temperatures. I^+ and I^- can be calculated by substituting the corresponding boundary temperatures T_1 and T_2 into the expression I_{eq}, where T_1 and T_2 are the bottom and top wall temperatures, respectively.

The local variation of temperature in terms of intensity with position can be obtained by plotting the normalized temperature $(I_{0eq}^*(\eta))$ versus η, as shown in Fig. 5.12. The $I_{0eq}^*(\eta)$ value varies between 0 and 1 since it cannot exceed the boundary value. At Kn = 0.1, a linear variation in temperature from one end to the other can be observed because Fourier's law applies to small Knudsen numbers. Therefore, the local equilibrium condition prevails at small Knudsen numbers. At Kn = 100, a perfectly flat line can be observed because there is no scattering between phonons. At very high Knudsen numbers approaching infinity, the transport phenomenon is called ballistic transport, where the phonons from one boundary directly strike the other boundary where the boundary temperatures are T_1 and T_2, respectively. The phonon comes with an intensity corresponding to T_1 and directly strikes the other surface, which results in a shock because the second surface is at temperature T_2. Therefore, the phonon does not adjust to the transition from T_1 to T_2, which results in a discontinuity at the boundaries. Similarly, the phonon from the second surface which is at temperature T_2 strikes the first surface which is at T_1, but it cannot adjust to the transition in temperature between T_2 and T_1. Therefore, the temperature jump can be observed at the boundary. Researchers often refer to this phenomenon as a "temperature slip," a fundamental characteristic of nanoscale transport perpendicular to thin films.

However, these discontinuities cannot be observed at the macroscale because the collisions of phonons are sufficient so that they can transfer all the information from one end to the other. In the nanoscale subcontinuum regime, this kind of transport

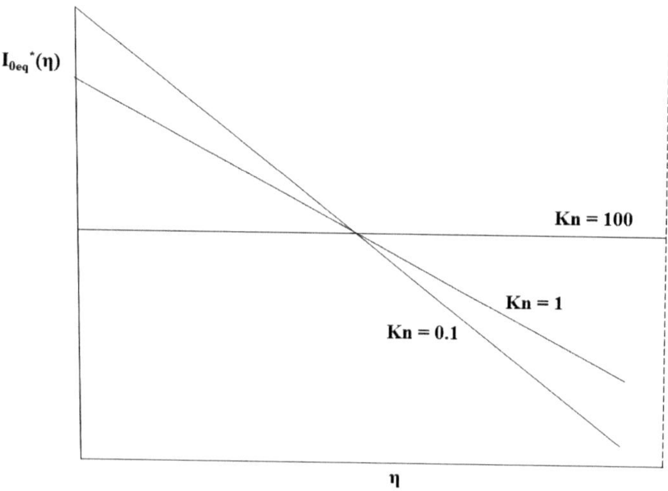

Fig. 5.12 Normalized temperature ($I_{0eq}^{*}(\eta)$) versus η

is not possible because the complete information between the energy carriers is not known. In other words, the emitted phonon intensity is different from the corresponding equilibrium value at $y = 0$. The equilibrium value is the summation of phonon intensity in all directions. In the nanoscale, the temperature is only an indicator of local energy. Therefore, at $y = 0$, $I_{eq} \neq I^{+}$. Similarly, at $y = d$, $I_{eq} \neq I^{-}$.

Finally, the primary understanding of the nanoscale energy transport is that the size effects become predominant at higher Knudsen numbers. The additional resistances at the boundaries are generated because of a "temperature slip," which results in a reduction of thermal conductivity. Therefore, effectively the nanoscale thermal connectivity is lower and decreases with increasing Knudsen numbers.

The transport phenomenon at nanoscales is explained very clearly in this chapter. The conservation equations, *viz.*, the Liouville equation, and the Boltzmann transport equation, which represents the position and momentum, are derived. The concept of relaxation time approximation is introduced to reduce the computational cost of solving the BTE. The classical continuum relations such as Fourier's law, Fick's law, Ohm's law, and Newton's shear stress law are derived using the BTE. Finally, the size effects in the nanoscale transport processes are described in detail.

Exercise Problems

1. The phonon–phonon scattering relaxation time in the intermediate range of temperature ($T < \theta_D$) can be approximated as follows:

$$\tau^{-1} = Ae^{-\theta_D/bT}.T^3\omega^2.$$

On the basis of Debye model, derive an expression for the thermal conductivity and discuss its dependence on temperature.

2. At high temperature, the phonon–phonon scattering relaxation time in a crystal is as follows: Where "a" is the order of distance between atoms and "m" is the atomic weight.

   (a) Prove that the high-temperature thermal conductivity is inversely proportional to T.
   (b) The thermal conductivity of silicon at $300\,\text{K}$ is $145\,\text{W/mK}$. Estimate its thermal conductivity at $400\,\text{K}$.

3. The electrical resistivity and thermal conductivity of gold at $300\,\text{K}$ are $3.1 \times 10^{-8}\,\omega\,\text{m}$ and $315\,\text{W/mK}$. Estimate the values of momentum and energy relaxation time, and the corresponding values of mean free path of electrons in gold.

4. The thermal conductivity of air is $0.025\,\text{W/mK}$. Estimate its dynamic viscosity.

5. Starting from the Cattaneo equation, derive the hyperbolic heat conduction equation. List the assumptions under which this equation is valid.

6. From the Boltzmann equation, by taking suitable moments derive the continuum continuity equation. What are the moments required to derive the momentum and energy equations (Navier–Stokes equations)?

7. Show that in metals, the electron contribution to thermal conductivity, k_e, is directly proportional to the electrical conductivity, σ, at a specified temperature, T, through the Lorentz number, $L = 2.45 \times 10^{-8}\ \text{W}\ K^{-2}$ as $L = k_e/\sigma T$. This relationship is referred to as Wiedemann–Franz law.

8. Derive an expression for the classical size effect of thermal conductivity along a thin film, assuming that the boundaries scatter phonons partly diffusely and partly specularly, with p representing the fraction of specularly scattered phonons.

Chapter 6
Microscale and Nanoscale Transport in Single-Phase Fluids

6.1 Introduction to Microchannels

The microscale phenomenon is not different from the macroscale phenomenon, and it is mainly governed by the kind of fluid used. In the case of rarefied gases, continuum assumption is not valid. However, in other cases such as microchannels, where liquids are used, the continuum equations can be used to analyze the flow physics. In the microscale regime, an interesting physical phenomenon, which is entirely different from that in the macroscale, can be observed. This unusual phenomenon can be demonstrated through the study of microscale energy transport. Section 6.2 mainly focuses on gas flows in microchannels as the deviation from the continuum in the microscale is primarily for gases.

Microchannels are increasingly becoming popular devices for effective cooling applications such as electronic cooling. While the Nusselt number is constant and equal to 3.66 in the case of a fully developed laminar flow through tubes with the uniform surface temperature boundary condition, it is 4.66 for a uniform heat flux.

The expression for the Nusselt number is given by

$$Nu_D = \frac{hD}{k}.$$

(6.1)

From this expression we see that, even if the Nusselt number is constant, the heat transfer coefficient is inversely proportional to the channel diameter. Therefore, microchannels become predominant in the present scenario for cooling applications.

The plot of the heat transfer coefficient (Y-axis) against hydraulic diameter (X-axis) for air and water shows the effect of hydraulic diameter on heat transfer coefficient. Although the Nusselt number is the same for both fluids, since the thermal conductivity of water is higher than that of air, the heat transfer coefficient of water is higher than that of air. Figure 6.1 shows that, as the channel diameter decreases, the heat transfer coefficient increases by up to three orders of magnitude. On reducing

© The Author(s) 2025

A. Pattamatta and S. K. Das, *Fundamentals of Nano- and Microscale Heat Transport*,
https://doi.org/10.1007/978-3-031-89613-2_6

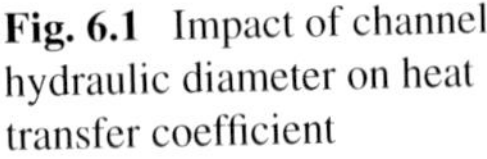

Fig. 6.1 Impact of channel hydraulic diameter on heat transfer coefficient

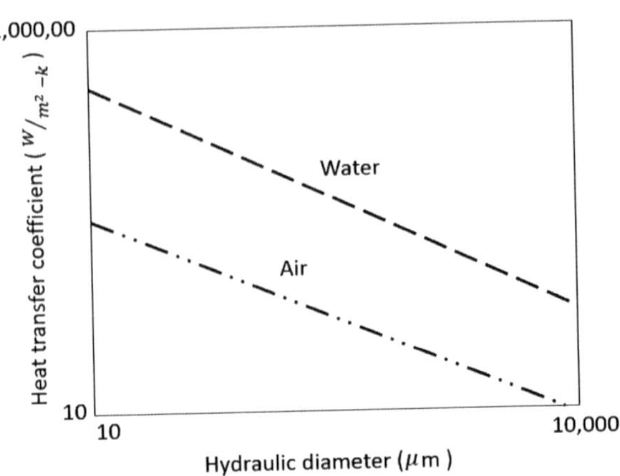

the channel diameter by 10 times, a 10-fold increase in the heat transfer coefficient can be observed.

Microchannels are very advantageous when the surface area is very small and, at the same time, a substantial amount of heat needs to be dissipated. To remove such an enormous amount of heat flux and maintain the system at a fixed temperature, the heat transfer coefficient must be very high. This can be achieved only using microchannels.

Therefore, from this simple analysis, one can conclude that microchannels are more efficient at improving the overall heat transfer rate compared to conventional channels. However, at the same time, the pressure drop penalty must be considered. In microchannels, a larger amount of pumping power is required to pump the same amount of liquid to maintain the same Reynolds number at the inlet.

Therefore, the pressure drop increases upon reducing the size of the channel. This means that the pressure drop (ΔP) is inversely proportional to the diameter (d). Hence, microchannels are advantageous from a heat transfer point of view but highly disadvantageous from the fluid dynamics point of view. This can be explained by the plot of pressure drop *versus* hydraulic diameter, as shown in Fig. 6.2. The figure shows that the pressure drop increases by up to seven orders of magnitude upon the reduction of the channel diameter by three orders of magnitude.

6.2   Gas Flow in Microchannels

To understand gas flow in microchannels, the following basic definitions have to be understood carefully.

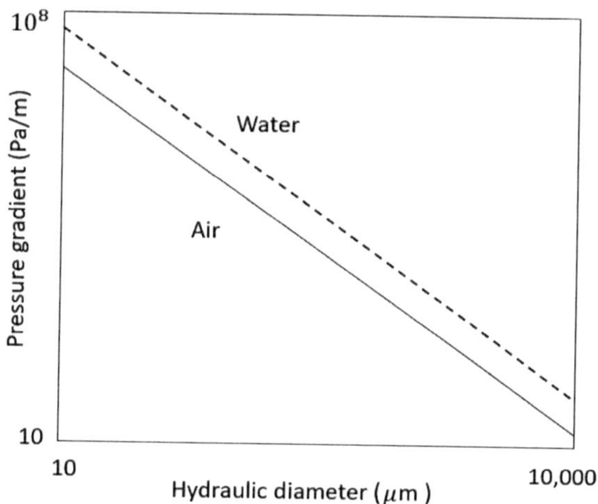

Fig. 6.2 Impact of channel hydraulic diameter on pressure drop

Properties

The properties are mainly associated with the continuum such as pressure, temperature, and density. Therefore, these are macroscopic manifestations of the molecular activity.

Continuum

It can be defined as the material that has a sufficiently large number of molecules in a given volume to give a unique value for properties. In other words, the properties are valid only in the continuum regime with a sufficiently large number of molecules to contribute to the defined averaged quantities such as pressure, temperature, and density.

The Validity of Continuum Assumption

The continuum assumption is not valid for large mean free paths.

Mean Free Path

The mean free path is the average distance traveled by the molecules between two successive collisions.

The expression for the mean free path at a given pressure and temperature is given by

$$\Lambda = \frac{k_B T}{\sqrt{2}\pi d^2 P}.$$ (6.2)

Kndsen Number

Knudsen number is defined as the ratio of the mean free path and the characteristic dimension of flow (channel hydraulic diameter).

$$Kn = \frac{\Lambda}{d}.$$ (6.3)

Upon non-dimensionalization of all these parameters, the kn out to be the ratio of the Mach number to the Reynolds number. Therefore, it is expressed as follows:

$$Kn = \sqrt{\frac{\pi \gamma}{2}} \frac{Ma}{Re_L}.$$ (6.4)

6.2.1   Flow Regimes Based on the Knudsen Number

Continuum Flow

If the Knudsen number is much smaller than 0.001 ($Kn \leq 0.001$), then the continuum approximation is valid. This is called the continuum regime. In this regime, thermodynamic equilibrium is also valid. Both velocity and temperature no-slip exist at the boundaries. Conventional Navier–Stokes equations can be solved to analyze the fluid flow and heat transfer problems.

Slip Flow

The continuum assumption is still valid for gas flow at moderate Knudsen numbers ($0.001 \leq Kn \leq 0.1$) as well as for liquid flow. Scattering of molecules with the boundaries takes place even at $Kn = 0.1$. The scattering between the gas molecules becomes more rarefied compared to the boundary scattering. Hence, nonequilibrium exists locally near the boundaries. The thermodynamic equilibrium concept is not valid throughout the fluid domain, which accounts for the slip effects at the boundaries. Therefore, this regime can be termed as the "slip flow regime."

Slip Mechanisms

Two types of slip exist: one accounts for the momentum and the other accounts for energy. Still, the Navier–Stokes equations can be solved to analyze the fluid flow problems in this regime. Instead of giving a no-slip boundary condition at the walls, the local nonequilibrium at the walls should be considered, but it cannot be accounted merely in the continuum equations. It is challenging to incorporate the transition from the local nonequilibrium near the wall to the continuum away from the wall using the slip boundary condition.

This is the reason why researchers resort to using the slip boundary condition models arbitrarily with empirical constants. However, it is not a very rigorous method.

Transition Flow

With the continuous increase in the Knudsen numbers, the continuum assumption utterly fails in the global scale even in the regions away from the wall.

The regime in which Kn varies between 0.1 and 10 ($0.1 < Kn < 10$) is called the transition flow regime. In this case, Burnett equations are used to analyze the flow physics. These equations are derived by expanding the Boltzmann transport equation to several terms and retaining some of the lower order terms. The lowest order term will be the Euler equation followed by Navier–Stokes equations and then the relatively higher order term will contain more information regarding higher Knudsen numbers. Velocity slips and temperature jumps exist in the transition regime.

Free Molecular Flow

If the Knudsen number is greater than 10 ($Kn > 10$), then inter-molecular collisions are negligible compared to the collision between the gas molecules and the wall. This is called the free molecular regime. Even the Burnett equations are not applicable in this case. This type of problem can only be analyzed by solving the Boltzmann transport equation.

Based on the value of the Knudsen number, the flow regime varies from the continuum flow to free molecular flow. Continuum approximation is valid till the slip flow regime. Beyond $Kn = 0.1$, there is a very thin separation between these regimes.

The Knudsen number is governed by two parameters: (1) the mean free path and (2) the characteristic length scale. The horizontal lines shown in Fig. 6.3 classify the transition from one regime to the other. The bottommost is the continuum regime and the topmost is the free molecular flow.

The Knudsen numbers are very small if the length scales are large and are of the order of 100 microns. The smaller length scales, which are of the order of a few

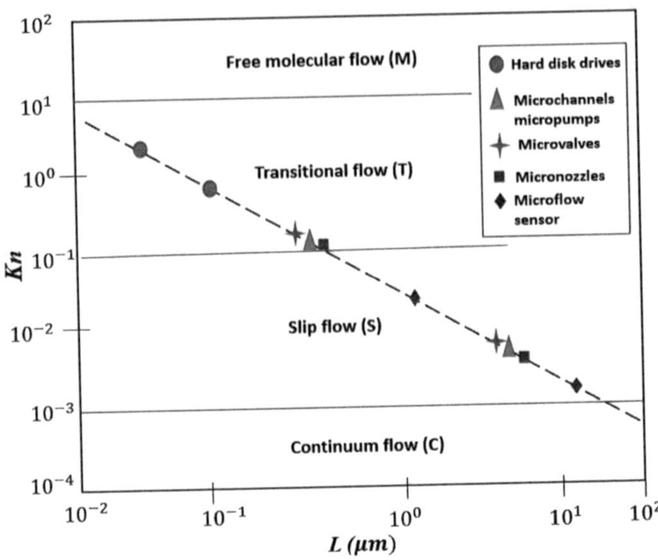

Fig. 6.3 Regimes of gas flow as a function of length scales

nanometers, correspond to the larger Knudsen numbers. In general, from the right toward the left, Knudsen number increases, as shown in Fig. 6.4.

Continuum equations are not applicable for the analysis of the complex situations, *viz.*, the air gap between the hard disk drive. In this case, the Navier–Stokes equations fail to predict the flow behavior. Therefore, Burnett equations with slip boundary conditions are used. In the case of microchannels with gas flow in the slip flow regime, continuum equations can be used, but they only account for the correction of local nonequilibrium at the wall. In liquid flows, the flow regime can be treated as a continuum. Therefore, the conventional continuum equations with a no-slip boundary condition at the wall correctly predict the flow behavior.

Therefore, perfect continuum assumptions can be used in microchannels. Most of the microchannels use liquids because the heat transfer coefficient is higher for liquids than for gases. The micronozzles fall in the transition or the slip flow regime. In this case, it is difficult to classify the kind of model that must be employed to analyze the flow behavior.

The Knudsen number is a function of the ratio of the Mach number to the Reynolds number. The flow regimes can also be classified based on the Mach number and Reynolds number by plotting the exit Mach number on the Y-axis and Reynolds number on the X-axis as shown in Fig. 6.4.

The Reynolds number in most of the microchannels does not exceed 1000 because of very small diameters. Even at very high velocities, at a Mach number of 0.31, the corresponding value of the Reynolds number will always be less than 1000. Therefore, in microchannels, there is no problem with analyzing turbulence. The

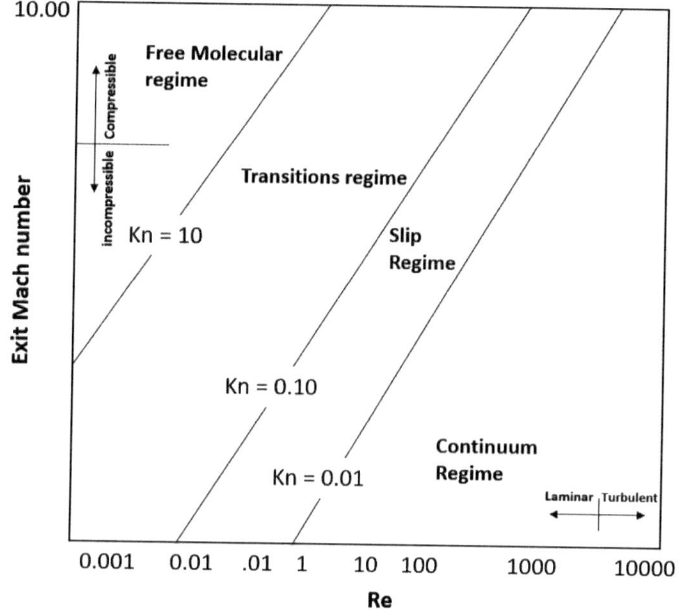

Fig. 6.4 Mach number as a function of Reynolds number

Reynolds number corresponding to $Kn \ll 0.01$ is approximately equal to 1000. Therefore, the flow is incompressible in such cases.

Draw a vertical line from $Re = 1000$ in such a way that it should intersect the graph. Draw a horizontal line toward the Y-axis from the intersecting point. From this, one can conclude that the Mach number corresponding to $Re = 1000$ is reasonably smaller enough to make an incompressible assumption. In the slip flow regime, at $Kn = 0.1$ and $Re = 100$, the Mach number tends to be higher than 1. This indicates that the flow becomes supersonic. The compressible phenomena can be observed even in the laminar regime. The increase in the exit Mach number depends on the value of the Knudsen number. The higher the Knudsen number, the higher is the Mach number value even for low Reynolds number.

For example, at $Kn = 10$, the flow becomes compressible even for a very low Reynolds number of 5×10^{-2}. In this case, the velocities are very high, i.e., higher than 120 m/sec corresponding to the Mach number of 0.3.

The Knudsen number of 10 is not a very common application even with gases. One of the typical applications is the hard disk drive. If $Kn = 1$, the corresponding Reynolds number may not be 10^{-2}, but it is close to 10^{-1}. The flow becomes compressible in this case. Therefore, one should be very careful with compressible gas flows, even for small Reynolds number for $Kn > 0.1$. The flow still becomes incompressible if $Kn < 0.01$. However, in the slip flow regime, if the operating Reynolds number is 1 or above ($Re \geq 1$), the flow becomes compressible. Therefore, most of the continuum models for the gas flow in microchannels can be solved with a compressible form of Navier–Stokes equation.

6.2.2 Flow and Heat Transfer in Microchannels

Macrochannel gas flows can be quantified by the Reynolds number (Re) and the Mach number (Ma). In the laminar regime, $Re < 2300$ and the corresponding Mach number is less than 0.3 ($Ma < 0.3$). Most likely, incompressible models will be used only in the case of macrochannels. In the case of microchannels, mostly even for very small Reynolds numbers, the flow becomes compressible. Models used to describe the gas flows in microchannels depend on rarefaction regimes. Rarefaction is the high Knudsen number effect.

Not only rarefaction effects but also compressibility effects will become important. Different models can assist us in the study of the flow physics in gas flow regimes based on Knudsen number (Kn), as shown in Fig. 6.5. A Knudsen number of 0 indicates the continuum regime with no viscosity. This means that it is entirely described by inviscid theory or Euler equations. In the case of finite Knudsen numbers in the range of 0.000001 with finite viscosity, viscous effects are significant. Therefore, Navier–Stokes equations with no-slip boundary condition can be solved to study the flow behavior. In the case of the slip flow regime ($Kn = 10^{-3} - 10^{-1}$), Navier–Stokes equations with the first-order slip can be used. There are different slip boundary conditions based on the Knudsen number.

The Navier–Stokes equations can be used to study the flow physics for higher Knudsen numbers, but the order of the slip at the wall has to be modified. For

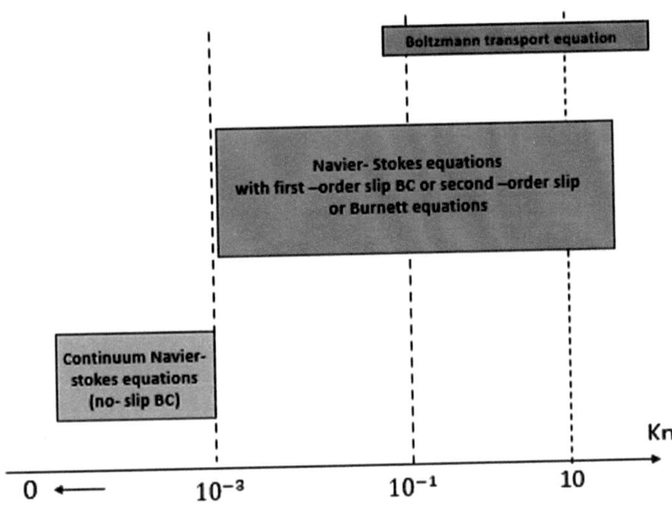

Fig. 6.5 Classification of flow regimes as a function of Knudsen number

$Kn > 0.1$, a model with a second-order slip boundary condition has been developed. The Navier–Stokes equations can be used primarily outside and near the wall. At the wall, a second-order slip can be used. However, there are problems with the numerical solution with a second-order slip. Therefore, the Burnett equations can be solved directly with the first-order slip boundary condition. Even for smaller length scales with Knudsen numbers of the order of 10 or above, the Boltzmann transport equation should be used. These are different models that have to be used to describe the flow behavior based on the Knudsen number (Kn).

The distinguishing factors of the fluid flow and heat transfer are compressibility and viscous dissipation effects. Viscous dissipation is governed by the ratio of the Eckert number to the Reynolds number. The Eckert number increases with the square of velocity. Viscous dissipation describes how the forces contribute to the work as heat dissipates in the fluid.

Rarefaction is nothing but low pressures in the case of high Knudsen number flows. Problems arise because of temperature-dependent properties, such as temperature jump. Large temperature gradients indicate higher local nonequilibrium. Slip velocity is another distinguishing factor. Thermal creep or thermal transpiration is the phenomenon in which the gas molecules move from one end to the other due to the temperature gradient. The surface forces become dominant at high Knudsen numbers.

The continuum approximation is valid up to $Kn = 0.1$ and the compressible Navier–Stokes equations can be used. The compressible Navier–Stokes equations in a coordinate-free form can be represented as follows:

Continuity equation:

$$\frac{\partial \rho}{\partial t} + \nabla.(\rho u) = 0. \tag{6.5}$$

Momentum equation:

$$\frac{\partial (\rho u)}{\partial t} + \nabla.\left(\rho u \otimes u) - \mu \left[\nabla \otimes u + (\nabla \otimes u)^T - \frac{2}{3}(\nabla.u)I \right] \right) + \nabla p = f. \tag{6.6}$$

Energy equation:

$$\frac{\partial E}{\partial t} + \nabla.\left((E + p)u - \mu \left[\nabla \otimes u + (\nabla \otimes u)^T - \frac{2}{3}(\nabla.u)I \right].u - k\nabla T \right) = f.u. \tag{6.7}$$

In the momentum and energy equations, the additional term $\frac{2}{3}(\nabla.u)$ is due to compressibility, and it will vanish for incompressible flows. Another term in the energy equation due to pressure work is given by $\nabla.(pu)$.

In the incompressible regime, the pressure work term $\nabla.(pu)$ is 0. These are the general Navier–Stokes equations for solving the gas flows in microchannels.

In the continuum regime, the no-slip boundary condition has to be applied. However, as the Knudsen number increases, there is a region near the wall where the gas molecules are out of equilibrium with the wall itself. Therefore, a temperature jump at the wall can be observed. Similarly, there is also a velocity jump becoming significant at higher Knudsen numbers. Therefore, the subcontinuum phenomena can be accounted in the continuum by introducing the wall slip boundary condition empirically.

There are two kinds of slips: velocity slip and temperature jump. A velocity slip exists in the momentum transport. In the conventional flow between two parallel plates, a parabolic velocity profile can be observed. The magnitude of velocity is zero at the wall. However, on increasing the Knudsen number to 0.1, the velocity profile is slightly different since a finite amount of velocity exists at the wall. This is called the velocity slip, which is similar to the temperature jump in the nanoscale transport phenomenon.

The mean free path is large for large Knudsen numbers. Therefore, gas molecules are likely to collide with the wall and travel with some momentum and energy. When they encounter the wall, a jump can be observed. Therefore, they suddenly cannot adjust to the transition in velocity. Hence, the jump continues. This jump becomes sharp at very high Knudsen numbers. This can be observed even at $Kn = 0.1$. Therefore, although this is entirely a molecular phenomenon, incorporating the jump into a continuum framework poses a challenge. One of the earliest models to incorporate the jump into a continuum was reported by Kundt and Warbury in 1875.

The apparent slip length can be calculated by drawing the tangent to the velocity profile and extending inside the solid boundary, as shown in Fig. 6.6. This tangent line intersects the vertical line at a depth of ξ from the edge of the wall. This value of ξ is called the slip length. The intersecting point can be treated as a virtual origin. The slip length approaches 0 at very small Knudsen numbers less than 0.001. In this case, the apparent origin exists on the wall. For higher Knudsen numbers, the virtual origin shifts into the wall because of the slip effects. Therefore, the slip length ξ has a finite value and increases with the Knudsen number.

Kundt and Warbury (1875) pointed out that for rarefied gas flows, a slip occurs at the wall:

$$u_{slip} = u_s - u_{wall} = \xi \left. \frac{\partial u_s}{\partial \eta} \right|_w . \tag{6.8}$$

The slip velocity at the wall is equal to the slip length times the gradient of velocity at the wall, where ξ is the slip length or the coefficient of the slip.

The coefficient of the slip is an empirical parameter. However, no study has been conducted on the analytical approach to calculate the coefficient of the slip. Therefore, researchers performed the experiments for different fluid and solid combinations and calculated the slip length. They extrapolated the velocity profile and measured the slip length, which can be used in the numerical model. There is no rigorous theory for obtaining the slip length, but there are some empirical ways of determining it. Maxwell in 1879 developed some expressions for inflating the slip length.

The slip length can be expressed as a function of accommodation coefficient, σ, and is given by

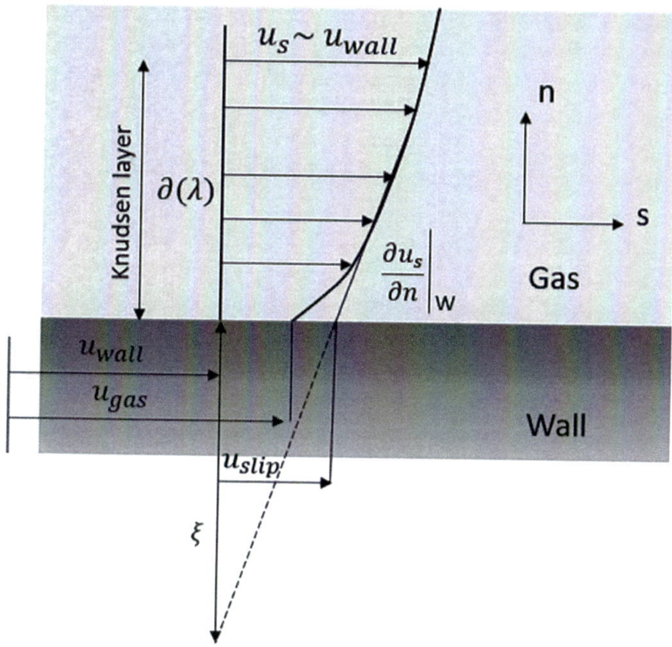

Fig. 6.6 Slip length calculation

$$u_{slip} = u_s - u_{wall} = \frac{2-\sigma}{\sigma}\Lambda\left.\frac{\partial u_s}{\partial n}\right|_W + \frac{3}{4}\frac{\mu}{\rho T}\left.\frac{\partial T}{\partial s}\right|_W. \tag{6.9}$$

The non-dimensional form of the above equation can be written as

$$u_s^* - u_{wall}^* = \frac{2-\sigma}{\sigma}Kn\left.\frac{\partial u_s^*}{\partial n^*}\right|_W + \frac{3}{4\gamma k_2^2}\frac{Kn^2 Re_0}{Ma_0^2}\left.\frac{\partial T^*}{\partial s^*}\right|_W, \tag{6.10}$$

where the second term is called thermal creep or transpiration phenomenon.

Maxwell proposed an idea to account for the cause of a slip in the microscopic continuum models. One of the most straightforward ideas is to draw an analogy from our reflection or scattering of phonons or electrons from the boundary by either specular or diffuse scattering.

In the case of diffuse scattering, Maxwell assumed that for every unit area, a fraction σ of molecules are absorbed or thermalized by the surface and afterward reemitted with a distribution function equal to the Maxwell–Boltzmann distribution function, with reemission occurring with a velocity of the gas at the wall temperature. The remaining $1-\sigma$ fraction of gas molecules will be simply reflected from the surface. The fraction σ is called the momentum accommodation coefficient.

The emitted gas molecules are thermalized based on the temperature of the boundary. Therefore, based on the Maxwell–Boltzmann distribution function, the momentum for the emitted gas molecules should be calculated. In the case of specular scattering, the momentum is perfectly conserved, but only the direction is changed. Therefore, for the same momentum, the reflected ray is equal to the incident ray of the gas molecules. The $1-\sigma$ fraction of gas molecules will be perfectly reflected in specular scattering. Therefore, these are two distinct and different ends of the spectrum. One is emitted equally in all the directions based on the boundary condition, called the diffused scattering phenomenon. The other is the specular scattering, which is entirely ignorant of the boundary condition. It is only emitted with the momentum of the incoming gas molecules on the boundary.

Therefore, the momentum accommodation coefficient signifies whether the momentum is perfectly conserved or not. In general, the momentum accommodation coefficient can be defined as the ratio of the difference between the incoming momentum (p_i) and the reflected momentum (p_r) of gas molecules to the difference between the incoming momentum and the momentum based on the Maxwell–Boltzmann distribution function (p_w). The expression for the momentum accommodation coefficient can be written in either the tangential or normal direction. $p = mv$ (momentum) and subscripts i and r represent the incident and the reflected ray of gas molecules and w refers to the Maxwell velocity distribution corresponding to the surface temperature T_w.

In diffused reflection, $p_r = p_w$ because the incoming gas molecules are absorbed, thermalized, and emitted based on the boundary conditions $f^+ = f_{eq}$. Therefore, in diffuse scattering, $\sigma = \sigma^1 = 1$. In specular scattering, the momentum accommodation coefficient is 0 because $p_r = p_i$ as the momentum is perfectly conserved, i.e., $\sigma = \sigma^1 = 0$.

Therefore, σ depends on the type of scattering, diffuse or specular, as shown in Fig. 6.7. Both are the possible boundary conditions that can be applied to solve the Boltzmann transport equation.

For air, the values of σ are quite high and toward the diffuse side. Therefore, σ lies between 0.87 and 1. For inert gases such as argon and nitrogen, the value of σ depends on the surface material. In the case of silicon as the manufacturing material for microchannels and nitrogen, argon, or carbon dioxide as the fluid, the values of σ vary between 0.75 and 0.85. They are always close to the diffuse scattering. In this case, if the interface is rough, the value of σ is close to 1 since most of the practical

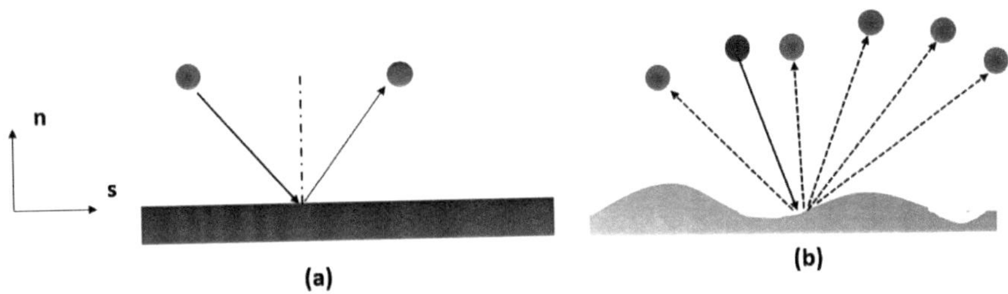

Fig. 6.7 **a** Specular reflection; **b** diffused reflection

surfaces have certain roughness. Therefore, the value of σ is always biased toward the diffuse scattering.

The σ value can be determined experimentally by calculating the slip length upon extrapolating the tangent to the velocity profile. Then, the slip length can be related to σ by the known expression.

Therefore, the first term represents one component of the velocity slip. In many experiments, researchers observed that the theoretical value of velocity slip does not match with the experimental value. There is also an extra component causing an additional slip at the wall and it cannot be explained by the known equations. Later on, it was understood that this is due to a phenomenon called thermal creep, in which an additional slip can be produced as a result of the temperature gradient.

In the case of a temperature gradient, gas molecules move from the lower temperature side to the higher temperature side in the direction of the applied temperature gradient. Therefore, in addition to the normal Maxwell slip, an additional slip due to thermal creep is observed. The second term represents the slip due to thermal creep. This is due to the fact that the temperature gradient along the wall is applied in the tangential direction, while the velocity gradient is applied in the normal direction to the wall.

Therefore, these two components when added together would adequately explain the slip velocity that can be measured experimentally. The slip boundary condition is popularly referred to as the Maxwell slip boundary. Some people refer to the Maxwell formulation as the Navier slip boundary. In the Navier slip boundary, some arbitrary values of slip length, such as 0.2, 0.5, and 0.7, can be given. However, the slip length in terms of accommodation coefficients can be related only by the Maxwell slip boundary condition.

Temperature Jump

There is a temperature discontinuity at the boundary at high Knudsen numbers, which can be termed as a "temperature jump."

A similar analogy has been drawn by replacing the momentum accommodation coefficient σ with the thermal accommodation coefficient σ_T to account for the temperature jump. It is assumed that the gas molecules with some energy approach the boundary, and a fraction of the gas molecules is absorbed and reemitted based on the Maxwell–Boltzmann statistics and the other fraction will be reflected.

However, in this case, to distinguish this fraction from the momentum accommodation coefficient, the notation σ_T can be used. The temperature slip is the difference in the temperature of the fluid at the boundary and the wall temperature. This slip length is called the temperature slip length.

Similar to the Maxwell velocity slip boundary condition, the temperature slip can be expressed in terms of the energy or temperature accommodation coefficients.

Therefore, the temperature slip can be written as

$$T - T_{wall} = \frac{2 - \sigma_T}{\sigma_T} \frac{2\gamma}{\gamma + 1} \frac{k}{\mu c_v} \Lambda \left. \frac{\partial T}{\partial n} \right|_w . \tag{6.11}$$

This is a well-known condition called the Smoluchowski boundary condition. Smoluchowski experimentally confirmed that this equation could be used to match the theoretical value of temperature slip with the experimental value very well by assuming the values for σ_T. Therefore, this boundary condition can be called the Smoluchowski slip boundary, whereas the velocity slip can be referred to as Maxwell slip boundary. On non-dimensionalizing the coordinate with respect to the characteristic length scale by introducing a new normal vector n^*, we obtain

$$T^* - T^*_{wall} = \frac{2 - \sigma_T}{\sigma_T} \frac{2\gamma}{\gamma + 1} \frac{Kn}{Pr} \left. \frac{\partial T^*}{\partial n^*} \right|_w , \tag{6.12}$$

where $n^* = \frac{n}{l}$ which is the normal coordinate, and similarly, the tangential coordinate is $s^* = \frac{s}{l}$. The thermal creep slip due to thermal creep also can be written in terms of $\frac{Kn^2 Re_0}{Ma_0^2}$.

Similarly, the temperature slip can be written in non-dimensional coordinates as the ratio $\frac{Kn}{Pr}$. Therefore, the microscale nonequilibrium at the wall can be incorporated into the continuum equations using these slip approximations for velocity and temperature such that it is quite easy to solve the governing equations for some standard cases such as flow between two parallel plates and flow in a duct. These are some simple cases for which the analytical solutions can be obtained. In the case of the slip regime, the no-slip boundary condition should be replaced with the appropriate Maxwell and Smoluchowski slip conditions.

The thermal accommodation coefficient in terms of energy can be expressed as

$$\sigma_T = \frac{\varepsilon_i - \varepsilon_r}{\varepsilon_i - \varepsilon_w}, \tag{6.13}$$

where ε is the average energy of a molecule and ε_w is the energy when the molecules are in thermal equilibrium with the wall.

In terms of temperature, it is given by

$$\sigma_T = \frac{T_i - T_r}{T_i - T_w}. \tag{6.14}$$

For specular reflection,

$$\varepsilon_r = \varepsilon_i, \ \sigma_T = 0.$$

For diffuse reflection,

$$\varepsilon_r = \varepsilon_w, \ \sigma_T = 1$$

$\sigma_T = 0.87 - 0.97$ for air–aluminum and air–steel systems.

$\sigma_T < 0.02$ for pure He gas and clean metallic surfaces.

The conservation equations for the compressible, Newtonian, and isotropic cases for which $Kn < 0.1$ are given as follows:

Continuity equation:

$$\frac{\partial \rho}{\partial t} + \frac{\partial}{\partial x_i}(\rho u_i) = 0. \tag{6.15}$$

Momentum equation:

$$\rho \left(\frac{\partial u_i}{\partial t} + \frac{\partial u_j u_i}{\partial x_j} \right) = \rho F - \frac{\partial p}{\partial x_i} + \frac{\partial}{\partial x_i} \left[\eta \left(\frac{\partial u_i}{\partial x_j} + \frac{\partial u_j}{\partial x_i} \right) + \Lambda \frac{\partial u_k}{\partial x_k} \delta_{ji} \right]. \tag{6.16}$$

Energy equation:

$$C_v \left(\frac{\partial T}{\partial t} + u_i \frac{\partial T}{\partial x_i} \right) = -P \frac{\partial u_i}{\partial x_i} + \phi + \frac{\partial}{\partial x_i} \left(k \frac{\partial T}{\partial x_i} \right), \tag{6.17}$$

where

$$\phi = \frac{1}{2} \eta \left(\frac{\partial u_i}{\partial x_j} + \frac{\partial u_j}{\partial x_i} \right)^2 + \Lambda \left(\frac{\partial u_k}{\partial x_k} \right)^2$$

Stress tensor:

$$\tau_{ji} = -p \delta_{ji} + \eta \left(\frac{\partial u_i}{\partial x_j} + \frac{\partial u_j}{\partial x_i} \right) + \Lambda \left(\frac{\partial u_k}{\partial x_k} \right) \delta_{ji}. \tag{6.18}$$

Heat flux vector:

$$q_i = -k \frac{\partial T}{\partial x_i}. \tag{6.19}$$

In this case, the compressible form of the Navier–Stokes equations along with the appropriate slip boundary conditions is solved. These equations can quite accurately predict most of the experimental results in the slip flow regime.

Consider a two-dimensional case of flow between two parallel plates, as shown in Fig. 6.8. This can be treated as flow-through microchannels. The separation distance between the two plates is $2h$. Two assumptions are made in this analysis. The first assumption is that the flow system is steady. Therefore, all the derivatives with respect to time can be assumed to be 0. Hence,

$$\frac{d\rho}{dt} = \frac{du}{dt} = \frac{dT}{dt}.$$

The second critical assumption is that the flow is fully developed. Therefore, there are no gradients of velocity in the z-direction. Also, this is a 2-D system. Hence, there

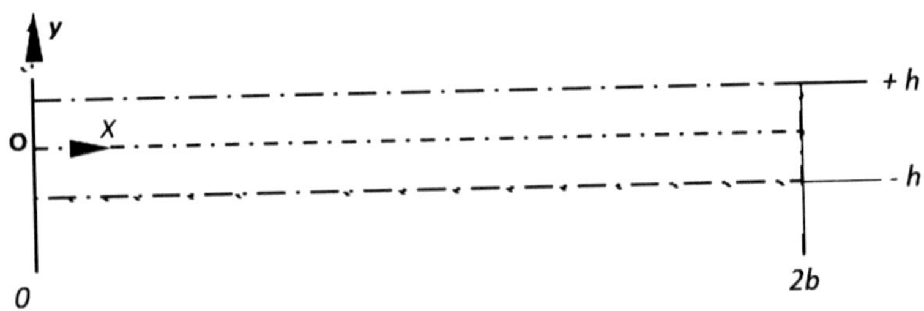

Fig. 6.8   Schematic of a 2-D Plane channel

are no gradients in the third direction. Therefore, there are no volumetric forces. This is a hydrodynamic problem.

Based on these assumptions, one can reduce the compressible Navier–Stokes equation to the following forms.

Only the viscous forces and the pressure gradient exist in this system. The viscous force exists on the top and bottom plates due to the diffusion of velocity in the vertical direction, which is equal to d^2u/dy^2. The viscous force is balanced by the pressure gradient dp/dz in the z-direction. Therefore, the momentum equation becomes

$$\frac{d^2u_z}{dy^2} = \frac{1}{\mu}\frac{dp}{dz}. \tag{6.20}$$

Therefore, the entire set of compressible equations reduce to a very simple form, which is similar to the incompressible equations for channel flow.

The boundary conditions are substituted to solve this equation. For this, the condition of symmetry at $y = 0$ is taken since the origin is fixed at the center of the microchannel, i.e.:

$$\left.\frac{du_z}{dy}\right|_{y=0}. \tag{6.21}$$

At $y = +h$, in the conventional channel case, the no-slip boundary condition ($u = 0$) has to be given. However, in the microchannel case, the Maxwell slip boundary condition is used, i.e.:

$$u_z|_{y=h} = -\frac{2-\sigma}{\sigma}\Lambda\left.\frac{du_z}{dy}\right|_{y=h}. \tag{6.22}$$

The general solution for the momentum equation is given by

$$u_z = \frac{1}{\mu}\frac{dp}{dz}\left(\frac{y^2}{2} + a_1 y + a_2\right). \tag{6.23}$$

This is a quadratic equation. The two boundary conditions at $y = 0$ and $y = h$ are substituted to obtain the constants a_1 and a_2.

In this case, the derivative is negative. Therefore, the velocity is maximum at the center and negative at $y = h$. Therefore, to obtain a positive value, du/dy should be negative. Otherwise, the negative value of velocity will be obtained, which is not physical.

Applying the given boundary conditions, the dimensionless form of the velocity distribution is given by

$$u_z^* = 1 - y_*^2 + 8\frac{2-\sigma}{\sigma}Kn = 1 - y_*^2 + 4\frac{2-\sigma}{\sigma}Kn^1, \tag{6.24}$$

where $y^* = \frac{y}{h}$, $u_z^* = \frac{u_z}{u_{z0}}$ and u_{z0} is the value of u_z at $y = 0$, $Kn = 0$. Substituting $Kn = 0$, the velocity profile can be obtained as

$$u_z^* = 1 - y_*^2\big|_{y=0}, \tag{6.25}$$

which is the maximum value at the central line corresponding to $Kn = 0$ at $y = 0$.

Therefore, the conventional parabolic profile with maximum velocity at the central line can be obtained. From this, the mean velocity at a given cross section can be evaluated by integrating the velocity profile from $-h$ to h. Therefore, the final expression for the value of the magnitude of mean velocity is given by

$$\bar{u}_z^* = \frac{\bar{u}_z}{u_{z0}} = \frac{2}{3} + 8\frac{2-\sigma}{\sigma}Kn = \frac{2}{3} + 4\frac{2-\sigma}{\sigma}Kn^1, \tag{6.26}$$

where $Kn^1 = 2Kn$. Therefore, the mean velocity increases with the Knudsen number. Therefore, for a given value of pressure drop, as the Knudsen number increases, the average velocity increases because of the higher slip. This means that an increased mass flow rate can be obtained for the same value of pressure drop.

The Fanning friction factor in terms of wall shear stress is given by

$$C_f = \frac{\bar{\tau}_w}{\frac{1}{2}\rho\bar{u}_z^2}. \tag{6.27}$$

The friction factor defined in terms of pressure drop is called the Darcy friction factor. Therefore, the relationship for Darcy friction factor in terms of Reynolds number for a circular macroduct is given by

$$f = \frac{64}{Re}. \tag{6.28}$$

The product of the friction factor and Reynolds number is called the Poiseuille number. In the case of macrochannel circular ducts, the Poiseuille number is 64. In the case of microchannel ducts with the slip boundary condition, the Poiseuille number is a function of Knudsen number. By force balance,

$$\bar{\tau}_w \, P dz = -A dp.$$

(6.29)

Solving and rearranging the terms

$$P_{0_{NS1,plan}} = \frac{24}{1 + 12\frac{2-\sigma}{\sigma} Kn} = \frac{24}{1 + 6\frac{2-\sigma}{\sigma} Kn^1},$$

(6.30)

where NS1 is the Navier–Stokes equation with a first-order slip flow boundary condition.

Therefore, the Poiseuille number decreases with the increase in Knudsen number. It can be inferred from the above equation that Poiseuille number is less than 24, which shows that the slip at the wall reduces the friction factor for a given pressure gradient and thus increases the flow rate.

When the Knudsen number tends to 0, the Poiseuille number approaches 24 and 36 in the case of channel and duct flows, respectively. Therefore, when the system is operating at a particular flow rate, the pressure required to drive the flow in the microchannel is lower than that in the macrochannel. In other words, on applying the same pumping pressure gradient to the pump, a higher flow rate can be obtained. Hence, it is a beneficial effect of the gas flow in the slip regime of microchannels.

Therefore, the non-dimensional mass flow rate can be calculated by dividing the actual mass flow rate, calculated from the slip boundary condition, by the mass flow rate calculated from the no-slip boundary condition, i.e.:

$$\left(\dot{m}_{NS1,plan}\right)^* = \frac{\dot{m}_{NS1,plan}}{\dot{m}_{ns,plan}} = 1 + 24\frac{2-\sigma}{\sigma}\frac{Kn_0}{\Pi+1} = 1 + 12\frac{2-\sigma}{\sigma}\frac{Kn_0^1}{\Pi+1},$$

(6.31)

where Π is the pressure ratio, which is the ratio of the inlet and outlet pressures. If the Knudsen number is 0, the non-dimensional mass flow rate is equal to 1, and for larger Knudsen numbers, the mass flow rate is greater than 1. This indicates that, as the Knudsen number increases, the non-dimensional mass flow rate increases as well.

Plotting the reciprocal of the non-dimensional mass flow rate as a function of Knudsen number, the reciprocal of the mass flow rate decreases with the increase in Knudsen number, as shown in Fig. 6.9. Different values of mass flow rates can be obtained based on the value of the accommodation coefficients.

If $Kn = 0.1$, then continuum equations with slip models can be used. However, if Kn crosses 0.1, the slip models are not accurate enough to obtain the solution. In those cases, Burnett equations are used. However, in some situations, researchers still use the continuum equations, but they modify the slip from a first-order to a second-order slip. To incorporate the higher order slip boundary condition, the term d^2u/dy^2 is incorporated into the governing equations.

The first-order slip boundary condition gives accurate results up to $Kn = 0.1$ and it is not suitable for higher Knudsen numbers. Using the second-order slip, even for

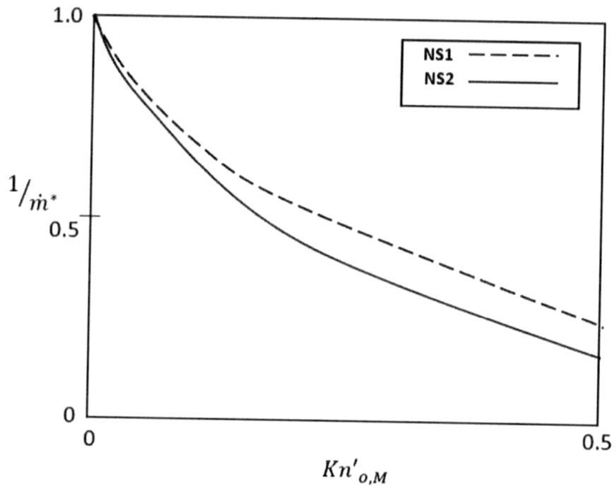

Fig. 6.9 Variation of the inverse flow rate with the Knudsen number based on the depth of microchannels

Knudsen numbers of up to 0.2, 0.3, good agreement with experimental results can be observed. Beyond $Kn = 0.3$, even the second-order slip will not be accurate.

For the problems that involve both fluid flow and heat transfer, the energy equation is solved along with the momentum equations. Therefore, the energy equation is similar to the macroscale parallel plate energy equation, and can be written as

$$\frac{k}{h^2}\frac{d^2 T}{dy^{*2}} = \rho u_z C_p \frac{\partial T}{\partial z}. \tag{6.32}$$

Heat transfer analysis in microchannels is straightforward similar to macrochannels. In this case, the only difference is that the temperature slip boundary condition has to be applied to the other wall. Therefore, at $y = 0$, the symmetry boundary condition and at $y = h$, the slip boundary condition should be applied.

The boundary conditions after neglecting the thermal creep are given by

$$\left.\frac{du_z}{dy^*}\right|_{y^*=0} = 0, \tag{6.33}$$

$$\left.u_z\right|_{y^*=1} = -\xi^* \left.\frac{du_z}{dy^*}\right|_{y^*=1}, \tag{6.34}$$

$$\left.\frac{d\theta^*}{dy^*}\right|_{y^*=-1} = 0, \tag{6.35}$$

$$\left.\theta^*\right|_{y^*=1} = -\xi^* \left.\frac{d\theta^*}{dy^*}\right|_{y^*=1}. \tag{6.36}$$

The dimensionless coefficient of the slip can be written as

$$\xi^* = \frac{\xi}{h} = \frac{\Lambda}{h}\left(\frac{2-\sigma}{\sigma}\right) = 4Kn\left(\frac{2-\sigma}{\sigma}\right), \tag{6.37}$$

$$\varsigma^* = \frac{\varsigma}{h} = \frac{\Lambda}{h}\left(\frac{2-\sigma_T}{\sigma_T}\right)\left(\frac{2\gamma}{\gamma+1}\right) \bigg/ Pr = 8\left(\frac{2-\sigma_T}{\sigma_T}\right)\left(\frac{\gamma}{\gamma+1}\right)\left(\frac{Kn}{Pr}\right). \tag{6.38}$$

The velocity distribution is given by

$$u_z = -\frac{h^2}{2\mu}\frac{dp}{dz}\left(1 - y^{*2} + 2\xi^*\right). \tag{6.39}$$

ς^* is a function of the accommodation coefficient, Knudsen number, and Prandtl number.

The non-dimensional temperature profile is given by

$$\theta^* = \varsigma^* + \frac{1}{1+3\xi^*}\left[\frac{1}{16}y^{*4} - \frac{3}{8}y^{*2}(1+2\xi^*) - \frac{1}{2}y^*(1+3\xi^*) + \left(\frac{13}{16}+\frac{9}{4}\xi^*\right)\right]. \tag{6.40}$$

The expression for the Nusselt number can be obtained from the non-dimensional temperature. If the Knudsen number is equal to 0, the value of the accommodation coefficient (ξ^*) becomes 0 at the wall. In the limiting case, where $Kn = 0$, the Nusselt number reaches the value of 140/17, which is the value of Nusselt number for macrochannels.

The concepts of velocity slip and temperature jump can be understood very well by plotting velocity and temperature profiles with and without slip conditions, as shown in Fig. 6.10. The temperature profile a is without any slip. On incorporating the slip, the temperature profile b shifts toward the right side. Comparison of these two profiles shows that the temperature gradient at the wall decreases from a to c. As a result, the Nusselt number and heat transfer coefficient also decrease.

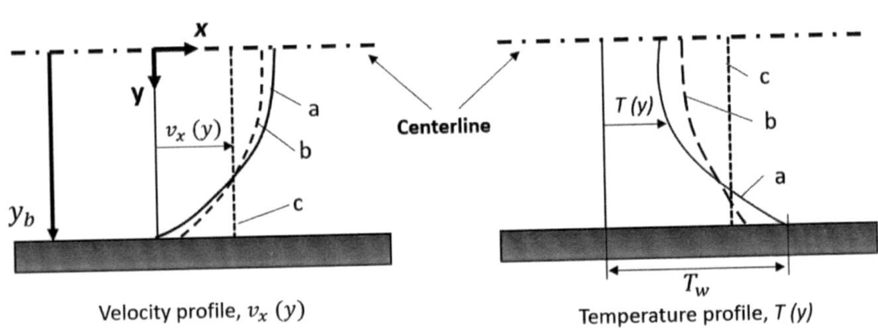

Fig. 6.10 Velocity and temperature profiles in internal flow across three regimes: **a** continuum regime, **b** velocity slip and temperature jump regime, and **c** free molecules regime

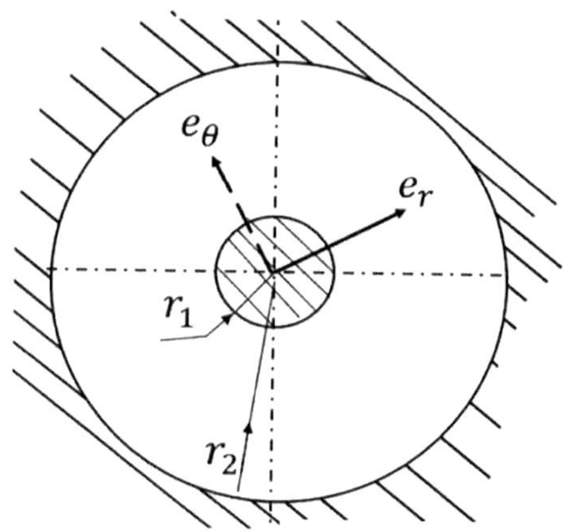

Fig. 6.11 Rarified gas flow in an annular duct

Therefore, due to the pressure drop, the convective heat transfer coefficient and Nusselt number also drop down. From the hydrodynamic point of view, it is very convenient to operate with slip flows, but from the heat transfer point of view, it is not beneficial. Therefore, if the operating Knudsen numbers are large, then the value of the Nusselt number decreases as given by the following expression:

$$Nu = \left(\frac{\xi_h^*}{4} + \frac{17 + 84\xi_h^* + 105\xi_h^{*2}}{140(1 + 3\xi_h^*)} \right)^{-1}. \tag{6.41}$$

For the macroscale case, the Nusselt number is equal to the classical value of 140/17.

Now, the gas flow in circular microtubes is briefly analyzed. A cross-sectional view of a circular microtube is shown in Fig. 6.11. The momentum equation for circular microtubes is given by

$$\frac{1}{r} \frac{d}{dr} \left(r \frac{du_z}{dr} \right) = \frac{1}{\mu} \frac{dp}{dz}. \tag{6.42}$$

The mean velocity distribution is given by

$$\bar{u}_z^* = \frac{1}{2} + 4 \frac{2 - \sigma}{\sigma} Kn. \tag{6.43}$$

This expression slightly changes for higher Knudsen numbers. The non-dimensional mass flow rate is given as

$$\left(\dot{m}_{NS1,circ} \right)^* = \frac{\dot{m}_{NS1,circ}}{\dot{m}_{ns,circ}} = 1 + 16 \frac{2 - \sigma}{\sigma} \frac{Kn_0}{\Pi + 1}. \tag{6.44}$$

The pressure distribution is given as

$$p^{*^2} + \beta_1 p^* + \beta_2 + \beta_3 z^* = 0, \qquad (6.45)$$

where

$$\beta_1 = 16\frac{2-\sigma}{\sigma}Kn_0$$

$\beta_3 = (\Pi - 1)(\Pi + 1 + \beta_1)$, and $\beta_2 = -1 - \beta_1 - \beta_3$. The Nusselt number is represented as

$$Nu = \left(\frac{\xi_r^*}{4} + \frac{11 + 64\xi_r^* + 96\xi_r^{*^2}}{48(1 + \xi_r^*)}\right)^{-1}. \qquad (6.46)$$

If the Knudsen number is 0, the value of Nusselt number becomes 48/11 (4.36), which is the case for a fully developed laminar flow with constant heat flux condition in the macroscale. In the case of microducts and microtubes, the Nusselt number is smaller than 4.36.

The expression for the Nusselt number for gas flow in 2-D microchannels is given by

$$Nu = \frac{140}{17 - 16\Lambda + (2/3)\Lambda^2 + 70\beta_T}. \qquad (6.47)$$

For 3-D cases, it becomes

$$Nu = \frac{48}{11 - 6\Lambda + \Lambda^2 + 48\beta_T}. \qquad (6.48)$$

A plot of Kn *versus* Nu shows that the Nusselt number is higher for parallel plates compared to that of circular tubes, as shown in Fig. 6.12. For different values of momentum and thermal accommodation coefficients, different values of Nusselt number can be observed. These expressions show that the Nusselt number is a function of the accommodation coefficient and Knudsen number. As the Knudsen number increases, the Nusselt number decreases.

The above analysis is valid up to $Kn = 0.1$. The continuum equations are still applicable at higher Knudsen numbers with second-order slip. However, the continuum assumption fails at very high Knudsen numbers. For Knudsen numbers greater than unity, Navier–Stokes, quasi-hydrodynamic, and quasi-gas-dynamic equations are invalid. For the early transition regimes, the slip flow model may still be valid if Navier–Stokes equations are replaced by Burnett equations. Therefore, the Boltzmann transport equation will be used for larger Knudsen numbers. However, solving the Boltzmann transport equation is computationally very expensive, especially in the case of three-dimensional flows. Alternatively, the Chapman–Enskog expansion can be used, which is a very classic way of deriving all the continuum equations starting from the Boltzmann equation. From the Chapman–Enskog expansion of the Boltzmann equation with Knudsen number as a small parameter, the form of the viscous stress tensor and heat flux vector starting from lower order terms to higher order terms can be written as follows:

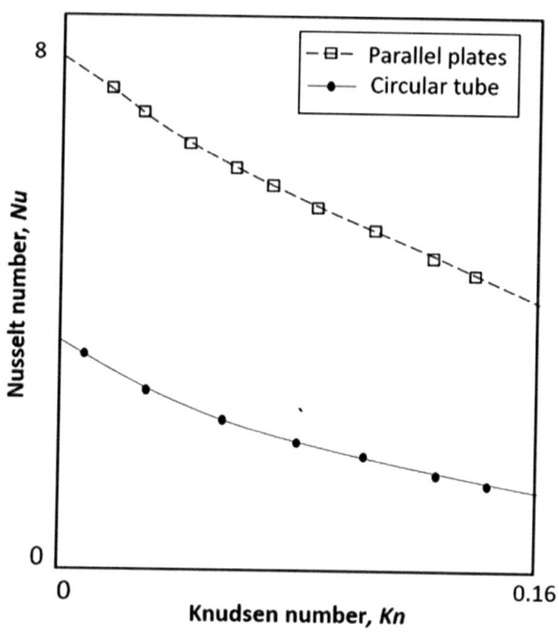

Fig. 6.12 Nusselt number as a function of Kn for air

$$\sigma_{EQ} = \sigma_{EQ}^{(0)} + \sigma_{EQ}^{(1)} + \sigma_{EQ}^{(2)} + \cdots + \sigma_{EQ}^{(i)} + \upsilon\left(Kn^{i+1}\right), \tag{6.49}$$

$$q_{EQ} = q_{EQ}^{(0)} + q_{EQ}^{(1)} + q_{EQ}^{(2)} + \cdots + q_{EQ}^{(i)} + \upsilon\left(Kn^{i+1}\right). \tag{6.50}$$

The process of removing all higher order terms and retaining only the lowest order terms in the Chapman–Enskog expansion is called the lowest order approximation, and the resulting equation is called the Euler equation, which therefore contains only zeroth-order terms. The terms that involve both viscous flux and heat flux are zero in Euler's equation. This indicates that the diffusion terms become zero.

Euler's equations can be written as

$$\sigma_{EQ} = \sigma_{EQ}^{(0)} = 0, \tag{6.51}$$

$$q_{EQ} = q_{EQ}^{(0)} = 0. \tag{6.52}$$

Incorporating both zeroth-order and first-order term results in the Navies–Stokes equations, which can be written as follows:

$$\sigma_{EQ} = \sigma_{EQ}^{(0)} + \sigma_{EQ}^{(1)} = 0 + \mu\left[\nabla \otimes u + (\nabla \otimes u)^T - \frac{2}{3}(\nabla.u)I\right], \tag{6.53}$$

$$q_{EQ} = q_{EQ}^{(0)} + q_{EQ}^{(1)} = 0 - k\nabla T. \tag{6.54}$$

Incorporating other higher order terms gives a better solution for higher Knudsen numbers. Therefore, researchers incorporated second-order terms for larger Knudsen numbers. As the Knudsen number increases, the second-order approximation should also be taken into account. The governing equations corresponding to the second-order approximation are called the Burnett equations.

Therefore, the equations that collectively include zeroth-order, first-order, and second-order terms are called the Burnett equations. When Knudsen numbers are greater than 0.3 or 0.4, inevitably Burnett equations are used to analyze the flow physics because these equations give a better solution compared to the continuum equations.

6.3 Liquid Flow in Microchannels

6.3.1 Liquid Transport in Minichannels and Microchannels

In liquid flow through microchannels, there is no problem with applying the no-slip boundary condition. The difference between the macroscale and microscale flow physics should be very well understood in the liquid flow. For this, the conventional macroscale liquid flow transport equations can be applied to the microscale, and the critical physics that can be analyzed in the microscale has to be emphasized.

Therefore, the Navier–Stokes equations can be applied to microchannels with liquid flow, but a few considerations should be taken into account, one of which is the effect of relative roughness.

The size of a microchannel is of the order of a few 100 microns. Hence, the effect of the wall roughness can be of the order of a few tens of microns, which can lead to different problems such as the transition from laminar to turbulent flow. Therefore, the relative roughness effect affects not only the basic friction factor and heat transfer coefficient but also the transition from the laminar to turbulent regime.

Certain empirical correlations for friction factor and Nusselt number are available for microchannels. However, whether these correlations are valid at the microscale directly should be verified. There are specific factors different from the macroscale to the microscale to be appropriately analyzed.

Therefore, one must investigate whether the correlations for macrochannels can be modified and applied suitably for the liquid flow in microchannels. In liquids, there are no problems associated with higher Knudsen numbers. The continuum assumption is valid, the flow is Newtonian, and there is no problem of velocity slip and temperature jump at the walls.

When examining flow through a uniform, one-dimensional, incompressible, circular pipe with a smooth surface and no roughness considerations, the analytical approach remains consistent across both macroscopic and microscopic scales. Regardless of the channel's diameter—whether it spans several centimeters or mere microns—a completely smooth wall maintains an identical friction factor and Nusselt number. Imagine a fluid segment with a length of dx within a pipe of diameter D. The equilibrium between frictional force and shear stress can be described as follows:

$$\frac{\pi}{4}d^2 dp = (\pi D dx)\,\tau_w.$$

(6.55)

Therefore, the shear stress and pressure gradient can be related as

$$\frac{dp}{dx} = \frac{4\tau_w}{D}.$$

(6.56)

From Newton's law of viscosity, the shear stress for Newtonian fluids is expressed as

$$\tau_w = \mu \left.\frac{du}{dy}\right|_w.$$

(6.57)

The Fanning friction factor f can be obtained from

$$\Delta p = \frac{2 f \rho u_m^2 L}{D},$$

(6.58)

where u_m is the mean velocity. The Darcy friction factor in terms of pressure gradient can be expressed as

$$f_{Darcy} = \frac{dp}{dx}\frac{D}{\frac{1}{2}\rho u_m^2 L}.$$

(6.59)

It is important to recognize that the Darcy friction factor is four times the value of the Fanning friction factor. For a circular cross section, the Fanning friction factor, when expressed in terms of the Reynolds number, is $16/Re$. Consequently, this means the Darcy friction factor equates to $64/Re$.

For a noncircular cross section, the hydraulic diameter is used. The hydraulic diameter (D_h) is given by

$$D_h = \frac{4A_c}{P_w},$$

(6.60)

where P_w is the wetted perimeter and A_c is the flow cross section.

In the case of a fully developed flow passing through a circular pipe, the friction factor f can be described as follows:

$$f = \frac{P_o}{Re},$$

(6.61)

where P_o is the Poiseuille number, which depends on flow geometry. For circular pipes,

$$P_o = f * Re = 16.$$

(6.62)

Duct shape			Nu_H	Nu_T	Po=f Re
Circular			4.36	3.66	16
Flat channel			8.24	7.54	24
Rectangular, Aspect ratio, b/a=		1	3.61	2.98	14.23
		4	5.33	4.44	18.23
		8	6.49	5.60	20.58
		∞	8.24	7.54	24.00
Ellipse, Major/Minor axis a/b =		1	4.36	3.66	16.00
		4	4.88	3.79	18.24
		8	5.09	3.72	19.15

Fig. 6.13 The Nusselt number and Poiseuille number values for different cross-sectional shapes and duct shapes

Therefore, P_o is constant for most of the fully developed duct flows, and depending on the cross-sectional shapes of ducts, different values of Poiseuille number can be obtained. Poiseuille number also depends on the aspect ratio in rectangular ducts.

Shah and London (1978) provided the following equation for a rectangular channel of sides a and b:

$$P_o = f\, Re = 24(1 - 1.3553\alpha_c + 1.9467\alpha_c^2 - 1.7012\alpha_c^3 + 0.9564\alpha_c^4 - 0.2537\alpha_c^5), \tag{6.63}$$

where $\alpha_c = \frac{a}{b}$.

For different cross-sectional shapes and duct shapes, Poiseuille number and different Nusselt number values for constant heat flux and constant wall temperature can be obtained, as shown in Fig. 6.13. With the increase in the aspect ratio, both the Poiseuille number and the heat transfer coefficient increase. Poiseuille number values for other shapes such as hexagon, triangle, and ellipse have been calculated. These values are applicable for both macrochannels and microchannels.

The hydrodynamic developing length (L_h) for macrochannels is given by

$$\frac{L_h}{D_h} = 0.05\, Re. \tag{6.64}$$

It is a non-dimensional length at which the boundary layers merge, and after which, no change in the velocity profile can be observed.

Therefore, beyond this length, $\frac{du}{dx} = 0$. Less than the developing length, the gradients in the axial direction are significant. Therefore, they cannot be neglected. To analyze this problem, the Navier–Stokes equations are solved by considering the inertial terms as they cannot be neglected in the developing region. The difference between the microchannel and macrochannel cases can be analyzed by keeping the channel diameters small for the same value of Poiseuille number. The pressure drop in a channel is given by

$$\Delta p = \frac{2 f_{\mathrm{app}} \rho u_m^2 x}{D_h}.$$

(6.65)

By keeping the friction factor constant, at $Re = 100$, and reducing the channel diameter by three orders of magnitude, i.e., from 1 mm to 1 micron, the pressure drop increases by several orders of magnitude. Hence, more pumping power is needed to overcome this pressure drop. Therefore, microchannels have relatively much smaller lengths compared to macrochannels.

In long channels, the developing length can be neglected. Developing lengths of several centimeters can be ignored. If the length is of the order of a few millimeters, it can be taken into consideration.

Therefore, one of the significant differences between microchannels and macrochannels is the effect of developing regions.

For microchannels, the expression for the pressure drop should be modified. The friction factor f is replaced with f_{app}. The apparent friction factor is taken into consideration because in microchannels the total pressure drop accounts for both developing and fully developed regions. For the developed regions, the Poiseuille number is constant for a given cross-sectional area, whereas for the developing region, empirical correlations are used to determine the Poiseuille number. f_{app} has two components: one accounts for pressure drop and is for a fully developed flow, and the other is for developing effects.

f_{app} is calculated by defining the incremental pressure (K), which is a function of x within the developing regime and becomes constant outside the developing region. Therefore, $K(x)$ can be defined as the difference between the apparent friction factor and the fully developed friction factor. It gives the local variation of the friction factor in the developing region. This is the reason the incremental pressure is a function of x. For x higher than the developing length (L_h), the incremental pressure attains the constant value because beyond this value the friction factor does not vary with respect to position. This constant value is known as Hagenbach's factor $K(\infty)$. The Hagenbach factor is a limiting value of the incremental pressure when x reaches the developing length. If the position is higher than the developing length, then $K(\infty)$ can be directly substituted and the apparent friction factor can be calculated. Therefore, the overall pressure drop can be defined in terms of f_{app}.

The pressure drop in terms of incremental pressure is given by

$$\Delta p = \frac{2(f_{\mathrm{app}} Re)\mu u_m x}{D_h^2} = \frac{2(f Re)\mu u_m x}{D_h^2} + K(x)\frac{\rho u_m^2}{2}.$$

(6.66)

For macrochannels, Hagenbach's factor can be ignored. ΔP can be calculated from the first term. However, for microchannels, the second term should also be included. Other standard correlations also exist to calculate the overall pressure drop as a function of x alone, without considering the effects of fully developed flow and Hagenbach's factor.

One such correlation of the frictional pressure drop for a circular duct is given by

$$\frac{\Delta P}{(1/2)\,\rho u_m^2} = 13.74\left(x^+\right)^{1/2} + \frac{1.25 + 64x^+ - 13.74\left(x^+\right)^{1/2}}{1 + 0.00021\left(x^+\right)^{-2}}, \qquad (6.67)$$

where the non-dimensional length x^+ is

$$x^+ = \frac{x/D_h}{Re}. \qquad (6.68)$$

Steinke and Kandlikar [90] obtained the curve fit for the Hagenbach's factor for rectangular channels as

$$K(\infty) = 0.6796 + 1.2197\alpha_c + 3.3089\alpha_c^2 - 9.5921\alpha_c^3 + 8.9089\alpha_c^4 - 2.9959\alpha_c^5, \qquad (6.69)$$

which is the most commonly used expression for rectangular channels. These are the empirical correlations from different experiments, which are applicable equally for both macrochannels and microchannels.

The apparent friction factor for rectangular ducts in the developing region for different aspect ratios has been plotted in Fig. 6.14. On plotting $f_{app}Re$ *versus* x^+ for different aspect ratios, $f_{app}Re$ increases, from the left to the right. For different aspect ratios, the value of K becomes entirely different, and in the developing region, $K(x)$ becomes significant. However, at $x < L_D$ (developing length), the effect of Hagenbach's factor can be ignored, and the fully developed friction factor becomes significant in that region. The curves for different aspect ratios coincide with each other in that zone.

For turbulent flow, Blasius developed the most straightforward correlation to estimate the friction factor, which is given by

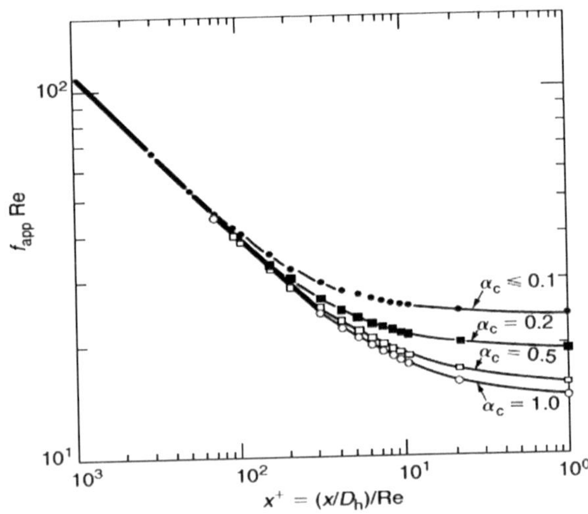

Fig. 6.14 Apparent friction factor for rectangular ducts in the developing region for different aspect ratios [81]

$$f = 0.0791 Re^{-0.25}. \tag{6.70}$$

Phillips developed an expression for the turbulent flow where both developing and developed regions are accounted. He presented the Fanning friction factor for circular tubes as a function of Re as follows:

$$f = A Re^{B}, \tag{6.71}$$

where

$$A = 0.09290 + \frac{1.01612}{x/D_h}$$

and

$$B = -0.26800 - \frac{0.32930}{x/D_h}.$$

For rectangular geometries, Re is replaced by the laminar equivalent Reynolds number:

$$Re^* = \frac{\rho u_m D_{le}}{\mu} = \frac{\rho u_m \left[\left(\frac{2}{3}\right) + \left(\frac{11}{24}\right)\frac{1}{\alpha_c}\left(2 - \frac{1}{\alpha_c}\right)\right] D_h}{\mu}, \tag{6.72}$$

where D_{le} is the laminar equivalent diameter.

Many microchannels can be machined on a surface. The number of channels could be 5, 10, 20, and 100. A feeder mechanism should be employed in the system, which supplies the liquid to all the channels through inlet manifold and collects the liquid through the outlet manifold on the other side.

To measure the pressure drop between the inlet and outlet manifolds, pressure tappings are connected to a differential pressure transmitter, which gives the overall pressure drop of the microchannel. In this case, apart from the developing region and frictional losses, entrance and exit losses ensue. Bend losses also exist at the entrance since the flow must turn 90°. The length of the inlet and outlet plenums depends on the total number of microchannels. Hence, the length could be significantly large. Therefore, in microchannels, the pressure losses within the plenum become significant. In microchannels, the entrance, exit, and bend losses can be neglected. Figure 6.15 shows a schematic representation of the experiments for pressure drop measurements in microchannels.

Therefore, the total pressure drop across the microchannel is given by

$$\Delta p = \frac{\rho u_m^2}{2}\left[\left(A_c/A_p\right)^2 (2K_{90}) + (K_c + K_e) + \frac{4 f_{app} L}{D_h}\right], \tag{6.73}$$

where A_c and A_p are the total channel area and the total plenum cross-sectional area, k is the loss coefficient at the 90° bends, K_c and K_e represent the contraction and expansion loss coefficients due to area changes, and f_{app} includes the combined effects of frictional losses and additional losses in developing a flow.

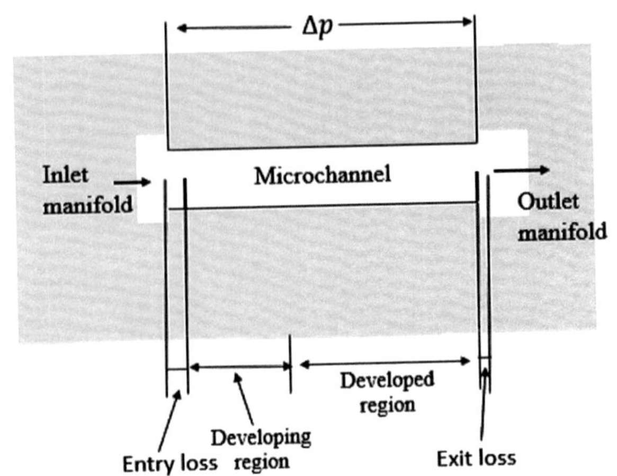

Fig. 6.15 Pressure drop in microchannels

On substituting f_{app} in terms of fully developed friction factor f and pressure drop defect $K(x)$, the expression for pressure drop between the inlet and outlet manifolds of a microchannel becomes

$$\Delta p = \frac{\rho u_m^2}{2}\left[(A_c/A_p)^2\,(2K_{90}) + (K_c + K_e) + \frac{4fL}{D_h} + K(x)\right]. \tag{6.74}$$

For $L > L_h$, $K(x)$ is replaced by Hagenbach's factor $K(\infty)$.

6.3.2 Effects of Roughness and Property Variation

Roughness is a very significant parameter in microchannels. The micro-structures of roughness can be observed with a microscope. Kandlikar [60] analyzed several parameters based on various roughness characterization schemes and defined a floor that acts as a foundation for roughness. The average height of the roughness can be estimated from the floor profile. The mean roughness height can be estimated from the dashed line, as shown in Fig. 6.16.

Parameters to Characterize Roughness

Average Maximum Profile Peak Height (R_{pm})

The average maximum profile peak height R_{pm} is the average distance from the individual highest points of the profile R_p, i to the mean line within the evaluation length. The mean line is representative of the standard average roughness value Ra.

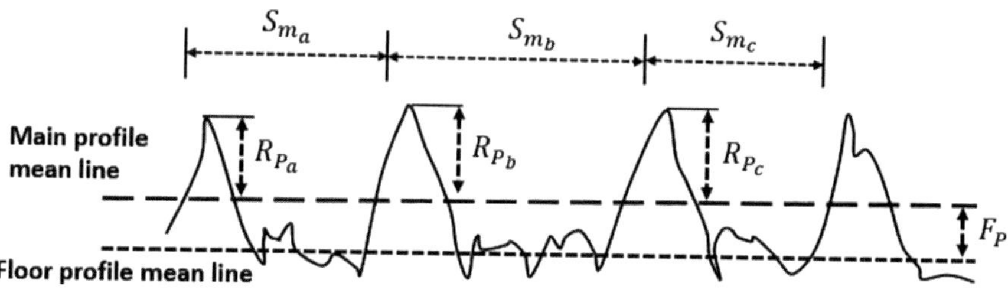

Fig. 6.16 Roughness parameters in microchannels

These values are expressed as R_{pa}, R_{pb}, and R_{pc}. The average of all these values results in the average maximum profile peak height R_{pm}.

Mean Spacing of Profile Irregularities (RS_m)

The mean spacing of profile irregularities (RS_m) refers to the mean value of the spacing between profile irregularities within the evaluation length. The irregularities of interest are the peaks, so this is equivalent to the pitch, i.e.:

$$RS_m = \frac{1}{n} \sum_{i=1}^{n} S_{m_i}. \tag{6.75}$$

Floor Distance to Mean Line (F_p)

It is the distance between the main profile mean line (determined by R_a) and the floor profile mean line. The floor profile is the portion of the main profile that lies below the main profile mean line.

Equivalent Roughness (ε)

The equivalent roughness is the summation of the floor distance to mean line F_p and the average maximum profile peak height (R_{pm}), i.e.:

$$\varepsilon = R_{pm} + F_p. \tag{6.76}$$

This equivalent roughness is used in many correlations where the friction factor or Nusselt number gets modified.

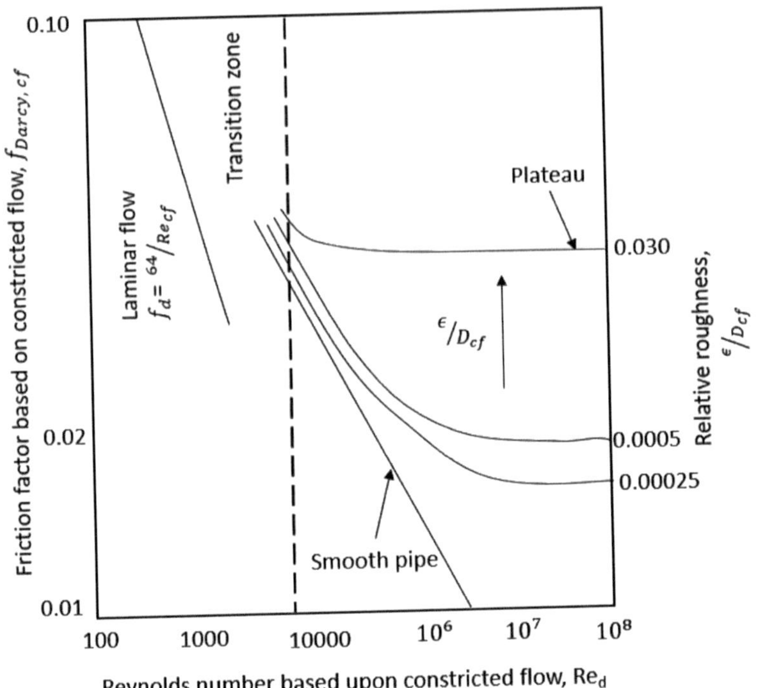

Fig. 6.17 Modified Moody diagram

The earlier conversation focused solely on smooth channels, where transferring the friction factor and Nusselt number from macrochannels to microchannels shows no disparity. Nonetheless, in practical channels, factors like equivalent roughness lead to higher friction factors and pressure losses than in smooth channels. To address these discrepancies, Kandlikar introduced a straightforward model named the constricted flow model. This model takes into account the decrease in cross-sectional area caused by protruding roughness elements and suggests utilizing the constricted flow area for friction factor calculations. As a result, a revised Moody diagram was developed, centering on the constricted diameter (D_{cf}), i.e.:

$$D_{cf} = D - 2\varepsilon. \tag{6.77}$$

The flow follows the constricted dimension rather than the original dimension. The modified Moody diagram can be obtained by replacing the original diameter d with D_{cf} since the original Moody chart was drawn based on the actual diameter d. On drawing the curves based on the constricted diameter (D_{cf}), for all microchannels, they matched very well with the modified Moody chart, as shown in Fig. 6.17.

The flow and geometrical parameters based on the constricted flow diameter are given by

$$\Delta p = \frac{2 f_{cf} \rho u_{m,cf}^2 L}{D_{h,cf}},\qquad(6.78)$$

where $u_{m,cf} = \frac{\dot{m}}{A_{cf}}$ and $Re_{cf} = \frac{\rho u_{m,cf} D_{h,cf}}{\mu}$.

Since the constricted flow area is smaller than the original area, the velocity increases. Therefore, the definition of Reynolds number is according to the velocity based on the constricted flow and the diameter is based on the constricted flow diameter.

The expression of the friction factor for a fully developed laminar flow and $0 < \frac{\varepsilon}{D_{cf}} < 0.15$ is given by

$$f_{cf} = \frac{Po}{Re_{cf}}.\qquad(6.79)$$

Therefore, the pressure drop is different in microchannels with roughness effects. The pressure drop gets modified by replacing the conventional velocity and diameter with the constricted flow diameter and the constricted velocity, while other parameters remain the same.

In the fully developed turbulent region for $0 < \frac{\varepsilon}{D_{cf}} < 0.03$, Halland (1983) obtained the relation for the friction factor as follows:

$$f_{cf} = \frac{f_{Darcy,cf}}{4} = \frac{1}{4}\left\{-18\log_{10}\left[\left(\frac{1}{3.7\left(D_{cf}/\varepsilon\right)+2}\right)^{1.11} + \frac{6.9}{Re_{cf}\left(\frac{D_{cf}}{D_{cf}+2\varepsilon}\right)}\right]\right\}^{-2}\left[\frac{1}{1+\frac{2\varepsilon}{D_{cf}}}\right]^{5}.$$

$$(6.80)$$

Schmitt and Kandlikar [86] studied the effect of relative roughness in an artificially and roughened rectangular microchannel. They concluded that for shorter pitches, with a pitch-to-roughness ratio of less than 5, the constricted model can predict the friction factor well. If the ratio is greater than 5, the smooth flow model should be used.

Smooth channels along with the roughness elements, such as aligned sawtooth, are shown in Fig. 6.18.

Microchannels can have roughness elements that are uniformly spaced at the top and the bottom called an aligned sawtooth. In this case, the constricted flow width can be calculated as $b_{cf} = b - 2 \times 72.9 \mu$m. The constricted flow width also decreases if there is an offset in which the top roughness is aligned separately from the bottom roughness. In this case, the constricted flow width can be calculated as $b_{cf} = b - 72.9\,\mu$m.

Using the expression for friction factor as $64/Re$, the considerable offset between the experiment and theory can be observed. Here, Re is based on the hydraulic diameter. The friction factor based on experiments matches very well with the theoretical values by using $f = 64/Re_{cf}$. Here, Re and Re_{cf} are calculated based on the hydraulic diameter and constricted flow diameter, respectively.

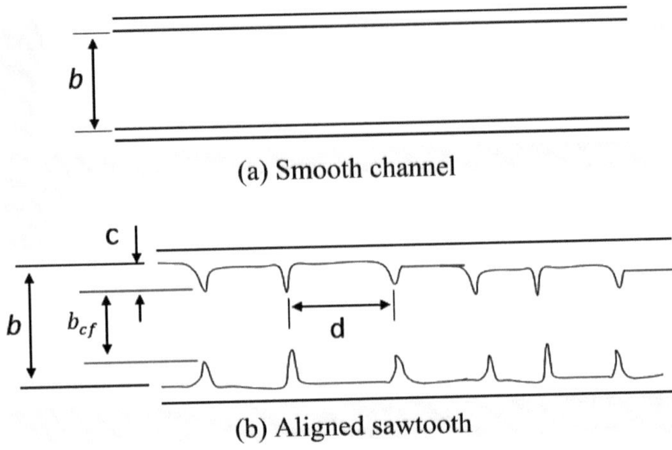

Fig. 6.18 Various roughness elements used **a** smooth channel; **b** aligned saw tooth

Roughness plays a significant role in the transition from laminar to turbulence regime. In a smooth surface, the transition takes place when the Reynolds number is around 2,300. If the channel has considerable roughness, the critical Reynolds number reduces.

The transition from the laminar to the turbulent region can be illustrated as follows:

For $0 < \frac{\varepsilon}{D_{h,cf}} \leq 0.08$,

$$Re_{t,cf} = 2300 - 18750 \left(\frac{\varepsilon}{D_{h,cf}} \right). \tag{6.81}$$

For $0.08 < \frac{\varepsilon}{D_{h,cf}} \leq 0.15$,

$$Re_{t,cf} = 800 - 3270 \left(\frac{\varepsilon}{D_{h,cf}} - 0.08 \right). \tag{6.82}$$

By substituting the roughness height as 0, the transition can be obtained as 2,300, which is the classical case, but for the values of $\frac{\varepsilon}{D} > 0$, $Re < 2300$. Therefore, the transition Reynolds number considerably decreases on increasing the roughness. On plotting the transition Reynolds number as a function of roughness height, as shown in Fig. 6.19, the experiments conducted for air and water correlated well. From these correlations, if the roughness is high, turbulence can be observed in microchannels even if the operating Reynolds number varies between 700 and 800.

In microchannels, for a fully developed laminar flow, the Nusselt number is constant and is 4.36 for the constant heat flux condition and 3.66 for the constant wall temperature condition. These values of the Nusselt number can be used for smooth

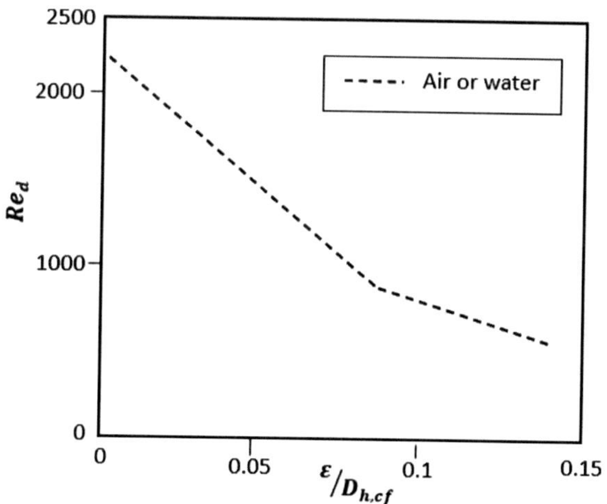

Fig. 6.19 Transition Reynolds number plotted as a function of non-dimensional roughness

microchannels in the laminar regime. Similarly, the turbulent correlations are also valid for microchannels without roughness.

Most of the microchannels do not have the circular cross section compared to macrochannels. The circular ducts are commonly manufactured using the machining process, while the rectangular ducts can be etched from a normal surface. The etching can be done by using photolithography. Usually, the etched microchannels will have a rectangular cross section with different aspect ratios. This is the reason why all these correlations are particularly emphasizing the use of different values of $\frac{a}{b}$ because most of these microchannels have a rectangular cross section with large aspect ratios.

The fully developed Nusselt number correlation as a function of α_c for the constant wall temperature case is given as follows:

$$Nu_T = 7.541 \left(1 - 2.610\alpha_c + 4.970\alpha_c^2 - 5.119\alpha_c^3 + 2.702\alpha_c^4 - 0.548\alpha_c^5\right).$$
(6.83)

The Nusselt number correlation for a constant circumferential wall temperature with a uniform heat flux is given by

$$Nu_{H1} = 8.235 \left(1 - 2.0421\alpha_c + 3.0853\alpha_c^2 - 2.4765\alpha_c^3 + 1.0578\alpha_c^4 - 0.1861\alpha_c^5\right).$$
(6.84)

Similarly, the Nusselt number correlation for the constant heat flux condition for both the circumferential and axial cases is given by

$$Nu_{H2} =$$

$$8.235 \left(1 - 10.6044\alpha_c + 61.1755\alpha_c^2 - 155.1803\alpha_c^3 + 176.9203\alpha_c^4 - 72.9236\alpha_c^5\right).$$
(6.85)

Therefore, the Nusselt number for the uniform heat flux case is always higher than that for the uniform wall temperature case. The same correlations can be used for smooth microchannels.

Thermally Developing Flows

The thermal entrance length can be obtained from

$$\frac{L_t}{D_h} = c\,Re\,Pr. \tag{6.86}$$

For circular tubes, $c = 0.05$, and for rectangular tubes, $c = 0.1$.

The local Nusselt number in the developing region of a circular tube is given by

$$Nu_x = 4.363 + 8.68\left(10^3 x^*\right)^{-0.506} e^{-41x^*}, \tag{6.87}$$

where $x^* = \dfrac{x/D_h}{Re\,Pr}$.

Fully Developed Turbulent Region

For $0.5 \leq Pr \leq 1.5$,

$$Nu = 0.0214\left[1 + \left(D_h/x\right)^{2/3}\right]\left[Re^{0.8} - 100\right]Pr^{0.4}. \tag{6.88}$$

For $1.5 \leq Pr \leq 500$,

$$Nu = 0.012\left[1 + \left(D_h/x\right)^{2/3}\right]\left[Re^{0.87} - 280\right]Pr^{0.4}. \tag{6.89}$$

For a fully developed turbulent region, Gnielinski correlation is given by

$$Nu_{Gn} = \frac{\left(f/8\right)(Re - 1000)\,Pr}{1 + 12.7\left(f/8\right)^{1/2}\left(Pr^{2/3} - 1\right)}, \tag{6.90}$$

$$Nu = Nu_{Gn}(1 + F), \tag{6.91}$$

where

$$F = C Re \left(1 - \left({^D/_{D_0}} \right)^2 \right),$$ (6.92)

$$f = (1.82 \log(Re) - 1.64)^{-2},$$ (6.93)

$C = 7.6 \times 10^{-5}$ and $D_0 = 1.164$. These are the standard correlations available for macrochannel heat transfer in the turbulent regime. These correlations not only are valid for circular cross sections but also can be applied to tubes with different cross-sectional areas and shapes.

In microchannels, the dimensions of the channel walls are comparable to the channel dimensions. Therefore, heat transfer in the walls cannot be neglected. This is called the axial conduction effect. Lin and Kandlikar (2012) considered the effect of axial conduction and derived the expression for the Nusselt number as follows:

$$\frac{Nu_{ko}}{Nu_{th}} = \frac{1}{1 + 4 \left(k_s A_{h,s} Nu_{th} / k_f A_f (Repr)^2 \right)},$$ (6.94)

where Nu_{ko} is the value of Nu obtained by neglecting the axial conduction effects, Nu_h is the value of Nu obtained with axial conduction effects, k_s and k_f are the thermal conductivities of solid and liquid, respectively, and $A_{h,s}$ and A_f are the heat conduction area for the wall and the cross-sectional area for the fluid flow, respectively.

Variable Property Effects

A variable property can be obtained by multiplying the known constant property with a correction factor. On considering the variation in viscosity, the viscosity ratios could be very high because the temperature at the wall could be entirely different from the bulk temperature. In this case, some simple correction factors are applied. Initially, the constant property is assumed and the friction factor and the Gnielinski number are calculated. Then, by multiplying the constant property with simple correction factors, the variable property is obtained. These variable property effects are common in both the macrochannel and microchannel cases.

In very high heat fluxes, the wall temperature is quite large compared to the bulk temperature. In those cases, the friction factors and Nusselt numbers need to be corrected because of non-uniformity in the correction of viscosity.

The friction factor is given by

$$\frac{f}{f_{cp}} = \left[\frac{\mu_w}{\mu_b} \right]^M.$$ (6.95)

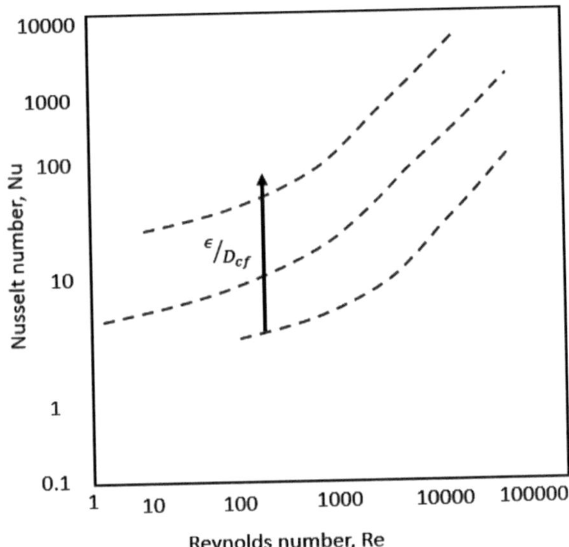

Fig. 6.20 *Nu versus Re* for single-phase liquid flow in microchannels and minichannels, $D_h = 50 - 600\,\mu\mathrm{m}$

The Nusselt number is given by

$$\frac{Nu}{Nu_{cp}} = \left[\frac{\mu_w}{\mu_b}\right]^N,\tag{6.96}$$

where cp refers to the constant property and the solution is obtained from appropriate equations. For a laminar flow, $M = 0.58$ and $N = -0.14$. For a turbulent flow, $M = 0.25$ and $N = -0.11$.

Some experiments have been conducted to understand the relative roughness effect. Kandlikar (2003) observed the effect of the relative roughness on the heat transfer performance by plotting *Nu versus Re* for three different roughness values (ε/d varies from 0.001 to 0.003) in the turbulent regime, as shown in Fig. 6.20.

Kandlikar concluded that for the same Reynolds number, the Nusselt number value increases with an increase in the relative roughness. Gnielinski's correlation can predict this effect well in the laminar regime.

In microchannels, roughness has a considerable effect on both the friction factor and Nusselt number. The constricted flow model can be applied to validate the friction factor and the Nusselt number with experimental results.

6.4 Electrokinetics

It is computationally very expensive to solve the Boltzmann transport equation for liquid flows in microchannels when most of the flow physics can be explained by continuum equations. Solving the Navier–Stokes equations is relatively easier compared to the Boltzmann transport equation.

Most of the correlations are equally valid in gas flows in both macroregimes and microregimes. The problem arises when the Knudsen number exceeds 0.1. If $Kn > 0.1$, then all the slip boundary conditions should be taken into consideration. However, for liquid flows, there is no problem. Therefore, researchers in this field have been attempting to explore and understand new phenomena at small scales. Electrokinetics is one among them. It is not a unique topic only related to the microscale; it can be applied to any surface or any length scale. However, it attains significance when the channel dimensions are smaller. Therefore, one can significantly use these electrokinetic forces to cause the motion.

Electrokinetics has a significant impact on microscale flows because microfluidics is the combination of microflows with electrokinetics. Microfluidics has become very important due to many applications in small scales. Electronic cooling could be one among them and it also finds applications in biological and medical fields. Therefore, people from these fields, despite having less expertise in fluid mechanics, want to build small-scale devices for biomedical applications.

The vital microfluidic functions required in various lab-on-a-chip devices are pumping, mixing, thermal cycling, dispensing, and separating. Pumping can be done in such a way that the pump should transport minimal volumes of fluid, i.e., in microliters, and no external mechanical pumping device is installed. The pumping device could be integrated within the same lab-on-a-chip device in such a way that the forces are used at the small scale to pump.

Mixing is another crucial aspect in microfluidics because the flow regime in microchannels is mostly laminar. In general, turbulence promotes the mixing process. Hence, it is challenging to achieve mixing in microchannels. Therefore, alternative ways of promoting the mixing process in microchannels are needed.

Separation is another important application of these microfluidic devices. One such application is to separate the platelets from plasma in blood. The fundamental idea is to build a portable handheld microfluidic lab-on-a-chip device that separates the plasma from the platelets.

Thus, microfluidics has gained popularity in the present scenario with much revenue compared to the other industries. The electrokinetic phenomena fall into two types: electro-osmosis and electrophoresis. In these processes, particles or fluids move under the influence of an applied electric field or electric potential. Electrokinetic phenomena can be further categorized as the streaming potential and the sedimentation potential.

Electro-osmosis

In electro-osomosis, the applied electric potential induces the motion of the fluid, which moves relative to the fixed plates or the fixed body walls.

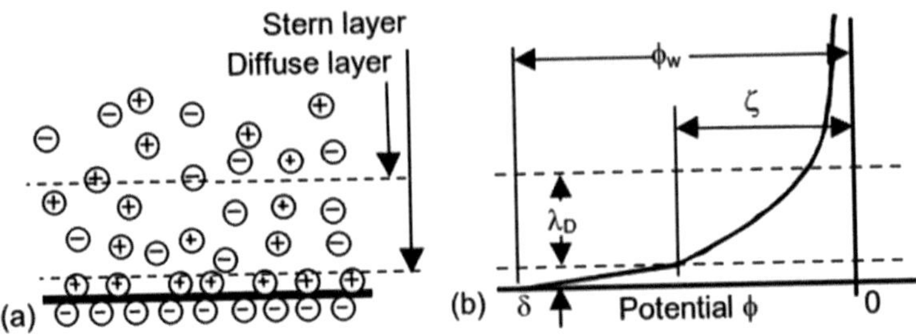

Fig. 6.21 **a** Schematic representation of Stern and diffuse layers; **b** wall and zeta potentials

Electrophoresis

In electrophoresis, the charged bodies or charged particles in a stationary fluid are accelerated by an external electric field. Therefore, the charged ions or metallic particles move relative to the fluid under the influence of the external electric field. On reversing this process, the external electric field can be induced by moving the charged particles or charged metallic surfaces.

6.4.1 Electric Double-Layer (EDL) Theory for Ionic Fluids

The fundamental physics behind electrokinetics can be explained by considering a polar fluid. When a wall or a boundary is brought into contact with the polar fluid, the surface of the wall acquires an electric charge.

In the given example, the surface is negatively charged, as shown in Fig. 6.21a. The excess negative charges at the surface are crucial at small scales. The base fluid has both positive and negative ions. For example, H^+ and OH^- exist in water. All positively charged ions migrate toward the negatively charged wall and form a first layer of positive charges called the Stern layer on the wall. The interaction potential between the positively charged ions and the negatively charged body is quite large.

However, there is a net negative charge on the wall, which cannot be compensated by the positive ions containing the Stern layer. Still, some more positive ions from the bulk fluid diffuse toward the wall. Therefore, a few positively charged ions diffuse toward the wall, at lengths which are much larger than the first layer and far away from the wall, there is no net charge accumulation. Hence, the fluid becomes electrostatically neutral. Therefore, another layer exists above the Stern layer called the diffuse layer. Outside the diffuse layer, the fluid becomes electrostatically neutral.

The electrostatic potential is highest at the Stern layer, decreases gradually toward the diffusion layer, and finally becomes equal to the bulk value. At very large distances, the electrostatic potential becomes 0.

Therefore, on applying the electric field, these charges move in the direction of the electric field. This phenomenon happens primarily at the diffuse layer because

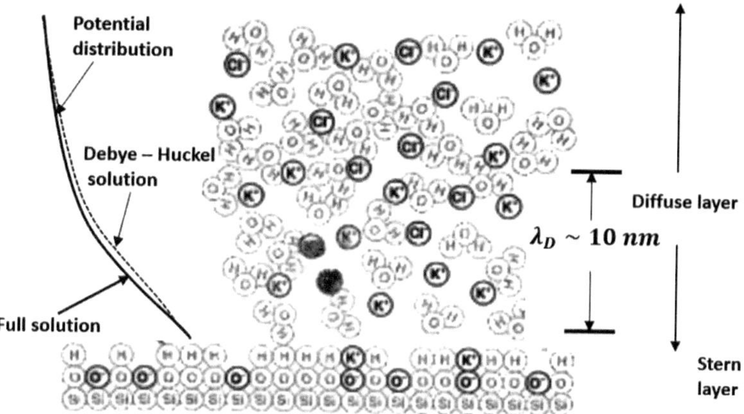

Fig. 6.22 The schematic of an electric double layer (EDL)

it is the thickest layer at which a maximum change in the electrostatic potential can be observed. Therefore, on applying the external electric field, all ions in the diffuse layer move when the Stern layer is fixed. This kind of motion can be interpreted as a Couette flow with the slip boundary condition.

This problem can be analyzed by solving for velocity in between the diffuse and Stern layers. The velocity on the diffuse layer can be given as a slip boundary condition. The solution in the outer region can be obtained by solving the Navier–Stokes equations without any external forces because they cannot induce any motion. It can be treated as the limiting case of a Couette flow in which both the top and bottom walls are moving with the same velocity. Therefore, a constant velocity is maintained between two diffuse layers at the top and the bottom. The velocity profile between these two plates can be obtained by averaging the wall velocities between the top and bottom plates if both walls move at the same velocity and in the same direction. In this case, the simplest solution for a Couette flow is the velocity at the edge of the electric double layer (EDL). Therefore, the velocity is uniform throughout the bulk.

The thicknesses of Stern and diffuse layers can be denoted by δ and λ_d, respectively. The potential at the edge of the Stern layer is ϕ and the potential for bulk fluid is ϕ_{bulk}. The potential difference between the bulk fluid and the edge of the Stern layer is called the double-layer potential or zeta potential (ξ), which is given as follows:

$$\xi = \phi - \phi_{bulk}. \tag{6.97}$$

A magnified view of the electric double layer is shown in Fig. 6.22. A silicon surface with a negative charge O^- is immersed in an aqueous solution of potassium chloride. All positively charged ions (H^+, K^+) are pulled close to the negatively charged O^- ions and form the Stern layer. A buildup of positively charged ions forms the diffuse layer over the Stern layer. Far away from the diffuse layer, the fluid becomes neutral.

Since the Stern layer is almost motionless, its thickness need not be calculated. Hence, the diffuse layer becomes significant. The thickness of the diffuse layer is generally of the order of a few tens of nanometers. On drawing the potential distribution, the potential at the wall gradually decreases to the bulk, as shown in Fig. 6.22. Therefore, the difference in potential between the bulk and any point is called the potential difference, which is 0 at the bulk and maximum at the wall.

The ϕ value specifies how the electrical potential at a specific point differs from the bulk value. The net charge density is calculated for each species of the ion. The local net charge density ρ_E as a function of the local potential is given by

$$\rho_E = \sum_i c_i Z_i F, \tag{6.98}$$

where c_i is the local concentration, Z_i is the valence number, and F is the Faraday constant, being 96,485 C/mol.

From Boltzmann's statistics, c_i can be written as

$$c_i = c_{i,\infty} \left(-\frac{Z_i F \varphi}{RT} \right). \tag{6.99}$$

Therefore, the Boltzmann solution is given by

$$\rho_E = \sum_i c_{i,\infty} Z_i F e^{\left(-\frac{Z_i F \varphi}{RT} \right)}. \tag{6.100}$$

In the actual Boltzmann statistics for energy, the term $\frac{E}{k_B T}$ appears in the numerator. Now, this term has to be rewritten. The energy is replaced by $Z_i F \varphi$, which gives the distribution of concentration from the wall to the bulk value at $c_{i,\infty}$.

The Poisson distribution establishes the relationship between the potential and the local net charge density:

$$\nabla^2 \phi = -\frac{\rho_E}{\varepsilon}. \tag{6.101}$$

Here, ε is permittivity. Therefore, the solution of the Poisson equation can be written as

$$\nabla^2 \varphi = -\frac{F}{\varepsilon} \sum_i c_{i,\infty} Z_i F e^{\left(-\frac{Z_i F \varphi}{RT} \right)}, \tag{6.102}$$

which is called the Poisson–Boltzmann equation. It establishes the relationship between the potential and free stream concentration of species. Normalizing this equation,

$$\nabla^{*^2} \varphi^* = -\frac{1}{2} \sum_i c^*_{i,\infty} Z_i F e^{(-Z_i \varphi^*)}, \tag{6.103}$$

where $\varphi^* = \frac{F\varphi}{RT}$. This is a non-dimensional electrostatic potential in terms of Faraday's constant R_T. Similarly, the length scales can be non-dimensionalized by using the Debye length λ_D, i.e.:

$$\lambda_D = \left(\frac{\varepsilon RT}{2F^2 l_c}\right)^{0.5}. \tag{6.104}$$

The Debye length gives the measure of the characteristic length over which the overpotential at a wall decays to the bulk. The one-dimensional form of the nonlinear Poisson–Boltzmann equation is

$$\frac{\partial^2 \varphi^*}{\partial y^{*2}} = -\frac{1}{2}\sum_i c^*_{i,\infty} Z_i F e^{(-Z_i\varphi^*)}. \tag{6.105}$$

If $Z_i\varphi^*$ is less than unity, the exponential term can be replaced with a first-order Taylor series expansion by setting

$$e^x = 1 + x. \tag{6.106}$$

Then the equation becomes

$$\frac{\partial^2 \varphi^*}{\partial y^{*2}} = -\frac{1}{2}\sum_i C^*_{i,\alpha} Z_i^2 \varphi^*. \tag{6.107}$$

Linearized Poisson–Boltzmann approximation can be obtained by using a Debye–Huckle approximation as follows:

$$\frac{\partial^2 \varphi^*}{\partial y^{*2}} = \varphi^*. \tag{6.108}$$

This is a simple second-order ordinary differential equation. Hence, two boundary conditions are required to solve this equation. Since the equation is in non-dimensional coordinates, the boundary conditions should be defined in the non-dimensional coordinates. Hence, the excess or overpotential at the wall is φ^*_0. Therefore, at $y^* = 0$, $\varphi^* = \varphi^*_0$.

At large values of y, the excess potential is equal to 0. Therefore, at $y^* = \infty$, $\varphi^* = 0$. Solving the equation with these two boundary conditions gives the solution in the form:

$$\varphi = \varphi_0 e^{-y/\lambda_D}. \tag{6.109}$$

Substituting φ into the expression for the concentration of the species (c) gives the solution to the one-dimensional Poisson–Boltzmann equation.

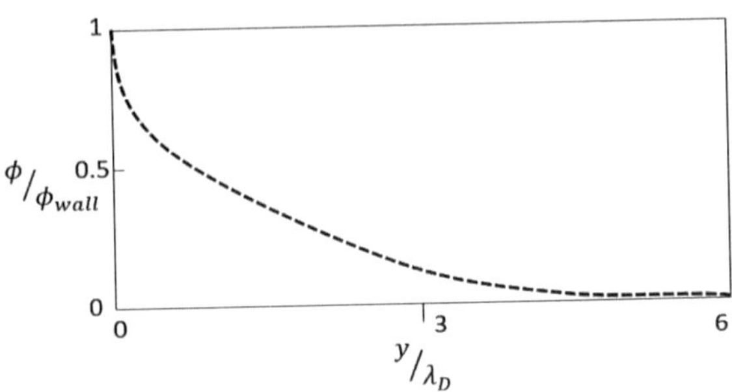

Fig. 6.23 Variation of the potential in a double layer with the distance

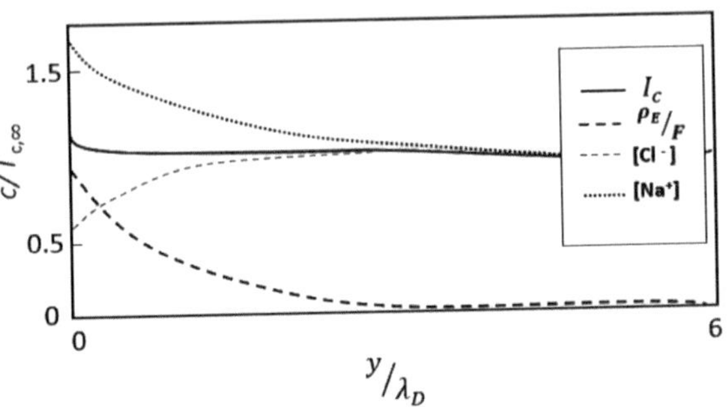

Fig. 6.24 Variation of the concentration in a double layer with the distance

This solution predicts the nature of the electrostatic potential precisely. On plotting this function φ/φ_{wall} with y/λ_D, exponential variation can be observed, as shown in Fig. 6.23.

For large values of y/λ_D, φ/φ_{wall} approaches 0. This is the variation of excess potential with distance. The variation of concentration with distance can be observed by plotting $dI_{c,\infty}$ versus y/λ_D, as shown in Fig. 6.24. The concentration of Na⁺ and Cl⁻ with respect to y can be plotted. In the case of a negatively charged wall, the concentration of positively charged ions is in excess close to the wall. Therefore, the concentration of Na⁺ is very high at the wall and then progressively decays and reaches the bulk concentration, which is equal to 1. For Cl⁻, the concentration is lower than 1 at the wall and then increases progressively and becomes equal to 1 in bulk.

When an electric field is applied to microchannels, bulk fluid motion can be observed. Applying an electric field along the positive x-direction generates a coulomb force ($f_{Coulomb}$). This force is the product of the local charge density (ρ_e) and the electric field and is given as

$$f_{coulomb} = \rho_e E. \tag{6.110}$$

In a flow between two parallel plates, as shown in Fig. 6.25, the variation of electrostatic potential can be observed up to Debye length, beyond which it becomes 0. The profile is symmetric about the central horizontal line.

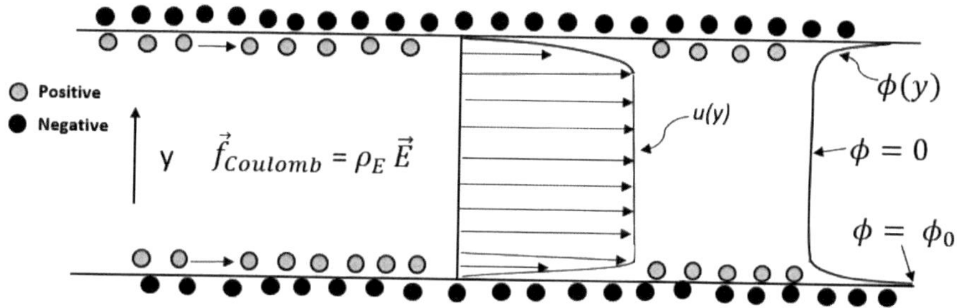

Fig. 6.25 Electro-osmotic flow between two parallel plates

The velocity of the motion is proportional to the applied electric field and depends on the material of the microchannel and the solution in contact with the wall. This motion is called electro-osmosis and is due to the influence of electrical forces on ions in the EDL.

The electric potential is high close to the wall and becomes 0 away from the wall since there is no force. Therefore, the electro-osmosis problem can be divided into two subproblems: inner and outer problems. In the case of the inner region, which is close to the wall, the velocity in the diffuse layer has to be solved and then one should consider the outer region, which is not directly influenced by the electric field, but the velocity boundary condition has to be applied.

Therefore, writing the Navier–Stokes equation for the thin region near the wall, we obtain

$$\rho\frac{\partial \overline{u}}{\partial t} + \rho \overline{u}\nabla\overline{u} = -\nabla p + \eta\nabla^2\overline{u} + \rho_E\overline{E_{\text{ext,wall}}}, \tag{6.111}$$

where $E_{\text{ext,wall}}$ is caused by the external power supply and is uniform within the EDL.

Here, the Debye length is of the order of tens of nanometers, and the mean free path is also very small for liquids. Therefore, $Kn \ll 0.1$. Hence, the continuum equations are valid. However, for the same length scales, continuum Navier–Stokes equations are not valid for gas flow because the mean free path is longer for gases. In the case of a steady, fully developed flow, both the time derivative and the convective acceleration terms become 0. Therefore, the entire left-hand-side term vanishes. On applying this analysis to a plate, where there is no pressure gradient, ∇p also becomes 0. Hence, only two forces exist: the viscous force and the Coulomb force. These two forces perfectly balance each other. Therefore, in the case of a fully developed steady-state solution for a flat plate with no pressure gradient along the x-direction, the N–S equations reduce to

$$0 = \eta\frac{\partial^2 u}{\partial y^2} + \rho_E\overline{E_{\text{ext,wall}}}. \tag{6.112}$$

The uniform permittivity Poisson equation can be written as

$$-\varepsilon\nabla^2\varphi = \rho_E. \tag{6.113}$$

Substituting ρ_E into the N–S equation reduces it to the form:

$$0 = \eta \frac{\partial^2 u}{\partial y^2} - \varepsilon \frac{\partial^2 \varphi}{\partial y^2} E_{\text{ext,wall}}. \tag{6.114}$$

Integrating from the wall to a point outside the EDL, the velocity profile as a function of φ and y can be obtained as follows:

$$\eta u = \varepsilon E_{\text{ext,wall}} \varphi + C_1 y + C_2. \tag{6.115}$$

On applying the no-slip boundary condition and forcing the velocity to be bounded at $y = 0$, the solution for the inner region is

$$u_{inner} = \frac{\varepsilon E_{\text{ext,wall}}}{\eta} (\varphi - \varphi_0). \tag{6.116}$$

Here, φ_0 is the potential at the wall, i.e., at $y = 0$. On plotting the velocity profile, a linear variation can be observed. The electric potential φ varies exponentially with y. The outer solution is driven by the solid wall with the velocity obtained for the inner solution. The EDL is a very thin layer close to the wall. Therefore, a velocity profile exists for the EDL for a few nanometers. At the edge of the EDL, the velocity obtained in the inner region can be applied as a boundary condition. Hence, the entire bulk flow is driven by the motion of the EDL. The edge of the EDL can be assumed to be a wall moving with the velocity that is equivalent to that obtained in the inner region. This flow situation can be considered as a Couette flow. Here, the bottom wall is at the edge of the EDL, which is in motion, and outside the edge of the EDL, another wall is also in motion. Therefore, the bulk value is the average value of the top and bottom walls. If both values are the same, the bulk velocity is the same as the boundary values.

The governing equation for the outer flow is given by

$$\rho \frac{\partial \overline{u}_{\text{out}}}{\partial t} + \rho \overline{u}_{\text{out}} \nabla \overline{u}_{\text{out}} = -\nabla p + \eta \nabla^2 \overline{u}_{\text{out}}. \tag{6.117}$$

In this case, the Navier–Stokes equations are to be solved without including the Coulomb force term as the net charge density is 0. The Navier–Stokes equation for the outer flow is solved with the interior boundary condition, which is the electro-osmotic slip velocity, i.e.:

$$\overline{u}_{\text{out}}(y = 0) = -\frac{\varepsilon \varphi_0 E_{\text{ext,wall}}}{\eta}. \tag{6.118}$$

On writing down the solution for the inner velocity (u_{inner}) at $y = \lambda_d$, i.e., at the edge of the EDL, the electric potential φ is approximately equal to 0. The velocity in the inner region turns out to be $-\frac{\varepsilon \varphi_0 E_{\text{ext,wall}}}{\eta}$. This has to be given as a slip boundary condition for the outer solution, which is called the Smoluchowski slip. In the case of gas flow for temperature, the temperature slip boundary condition is also called the Smoluchowski slip, whereas in the case of electrokinetics for velocity, it is called the Helmholtz–Smoluchowski slip boundary condition.

Therefore, to obtain the full solution for the outer region, conventional N–S equations should be solved with the Smoluchowski slip boundary condition at $y = 0$ for the limiting case, where $\nabla p = 0$. For the steady state, fully developed solution, the N–S equation becomes

$$\frac{d^2 u}{dy^2} = 0. \tag{6.119}$$

The linear profile can be obtained by solving this equation. Therefore, the entire electro-osmosis problem can be solved in two steps. In the first step, the solution for the inner region needs to be found out and used as a boundary condition for obtaining the solution for the outer region.

6.4.2 Electrokinetic Pumps

In electrokinetic pumps an electric field is applied along a capillary to generate flow and pressure. Unlike conventional mechanical pumps in which the flow can be driven purely by the pressure gradient, in these pumps, both an electro-osmotic-driven flow and a pressure-gradient-driven flow act simultaneously.

Therefore, on applying an electric field along a capillary, a strong electro-osmotic motion can be induced, which in turn generates the flow. Also, a pressure gradient drives the flow.

Consider a system in which an electric field $E = \frac{\nabla V}{L}$ is applied across an open rectangular microchannel with depth $2d$, width w, length L, and cross-sectional area $A = wd$. Assume that no pressure gradient exists and $\lambda_D << d << w << L$.

For a 2-D electro-osmotic flow, the volumetric flow rate can be obtained by integrating the wall velocity from $-d$ to $+d$ across the entire channel height, i.e.:

$$Q_{\text{EOF}} = W \int_{-d}^{d} u_{\text{wall}} dy. \tag{6.120}$$

This is the component obtained from the electro-osmotic process. Assume that only a pressure gradient is driving the flow and not electro-osmosis. In this case, the external electric field becomes 0 and the conventional solution for flow in a channel with a pressure gradient gives the flow rate as follows:

$$Q_{PDF} = W \int_{-d}^{d} \frac{1}{2\eta} \left(\frac{\partial p}{\partial x} \right) (d^2 - y^2)\, dy. \tag{6.121}$$

Therefore, this can be treated as a pressure-driven channel flow. In the case of a generic capillary, both electro-osmotic and pressure-driven flows co-exist. Therefore, the first term can be obtained by neglecting the pressure gradient completely, and the second term can be obtained by neglecting the electric field. However, in the actual case, both these terms govern the flow.

The total flow rate Q can be obtained by integrating and summing up both the expressions for Q_{EOF} and Q_{PDF}, i.e.:

$$Q = 2wu_{wall}d + \frac{2w}{3\eta} \left(-\frac{dp}{dx} \right) d^3. \tag{6.122}$$

Rearranging these terms, we obtain

$$Q = Q_{max} \left(\frac{\Delta p_{max} - \Delta p}{\Delta p_{max}} \right). \tag{6.123}$$

It is driven by both the slip velocity at the wall, which is the electro-osmotic solution for the EDL, and the pressure gradient. This is the mechanism of electrokinetic pumps.

Therefore, two limiting cases exist in electrokinetic pumps. In the first case, the pump works and a full flow without any pressure gradient exists due to electro-osmosis. In the second case, there is no flow, but the pressure gradient exists. This indicates that the exit has to be closed so that a maximum increase in pressure can happen. Therefore, there is no outlet for the fluid to flow.

The solutions can be obtained for these aforementioned extreme cases. In the case of pressure-driven flow, the condition $Q_{out} = 0$ has to be substituted into the Q_{PDF} and the relation for ∇p is obtained.

Therefore, integrating and setting the net flow rate equal to zero, we obtain

$$\frac{\Delta P}{\Delta V} = -\frac{3\mu_{E0}\eta}{d^2}. \tag{6.124}$$

This gives the expression for the maximum pressure rise as follows:

$$\Delta p_{max} = -\frac{3\mu_{E0}\eta \nabla V}{d^2}. \tag{6.125}$$

In the case of an open capillary with a maximum flow rate and no pressure gradient, the expression for the maximum flow rate can be obtained by substituting the pressure gradient as 0 into the expression for total flow rate:

$$Q_{max} = 2wd\mu_{EO}\frac{\nabla V}{L}. \qquad (6.126)$$

Now, the thermodynamic efficiency for a particular pump has to be defined. The thermodynamic efficiency has both electro-osmotic and pressure-driven contributions. It is the ratio of the pumping power generated by the pump to the total input power to the system. The total input power is the external electric field. Therefore, the thermodynamic efficiency is given by

$$\xi = \frac{\Delta p\, Q}{\Delta V I}. \qquad (6.127)$$

The pumping power that can be generated by the pump is equal to $\Delta p Q$. Here, $\Delta V I$ is the external electric field. The efficiency of electrokinetic pumps is generally less than 10%, but they can be used as an alternative to using a mechanical pump externally.

Electrophoresis

The electrophoresis phenomenon is an extension of electro-osmosis because, in the latter case, the walls are stationary and the fluid is in motion. In electrophoresis, the particles are in motion relative to the bulk fluid. Therefore, this phenomenon is termed as particle electrophoresis. The particles move due to the accumulation of charge around them under the influence of an external electric field. Therefore, it is assumed that the bulk fluid is not moving, but the charges are all concentrated around these particles, which move relative to the bulk fluid.

In many situations, these dispersing particles also have an electrostatic potential and electric double-layer formation occurs. On applying an electric field, the particles start migrating. This is a critical phenomenon in the case of nanofluids. The same theory of electro-osmosis for calculating the velocity can be applied so that the particle and the Debye layer move together. Therefore, no slip is present between the particle and the Debye layer. Hence, the velocity of the Debye layer is the same as that of the particle.

Due to the motion of these particles, a concentration gradient develops and the field is set up. In this case, the forces acting on the particles and the velocity of these particles can be analyzed similar to electro-osmosis by assuming the field to act in the opposite direction. This is because in the earlier case, the wall is stationary and the ions are moving because of the applied external field. In electrophoresis, the bulk fluid is stationary and the wall is moving.

Therefore, an electric field in the opposite direction has to be imposed. Hence, the same solution can be obtained as in electro-osmosis, but only the signs are reversed. In electrophoresis, viscous diffusion acts in the opposite direction to the applied Coulomb force. However, in electrophoresis, both act in the same direction so that the sign is positive.

The force exerted by the electric field is given by

$$\overline{F} = Ze\overline{E}. \tag{6.128}$$

In electro-osmosis, a fixed coordinate frame of reference is assumed such that the wall is fixed and the bulk fluid moves. In electrophoresis, the particle is moving and the bulk fluid is stationary. On fixing the coordinate to the particle, it becomes similar to electrophoresis, where the wall is stationary and the bulk fluid is moving, but the electric field is acting in the opposite direction. Hence, the same expression can be used for velocity by only replacing the potential mean with a negative sign.

Therefore, for the particles within EDL, electrophoretic velocity is given by

$$\overline{u_{EP,i}} = \mu_{EP,i}\overline{E}, \tag{6.129}$$

where

$$\mu_{EP} = -\mu_{E0} = \frac{\varepsilon\varphi_0}{\eta}.$$

This is a very simple approximation, which is particularly valid for a thin EDL. In this case where the radius of the particle or the diameter of the particle is much larger than the Debye length, the same velocity can be applied because the curvature appears to be very small for the bulk fluid.

Therefore, this can be considered as the case of a flow past a flat plate. In this case, the EDL is very thin and we can assume that both the EDL and the particle move with the same velocity.

In the case of a thicker EDL, the electric field itself will not be uniform because the Debye layer is quite thick and there is a variation in the velocity of the particles. A drag force exists between the particle and the EDL. Therefore, there is a slip effect between the EDL and the particle itself. Therefore, the velocity should be corrected using a multiplicative factor f, i.e.:

$$\overline{u_{EP,i}} = f\frac{\varepsilon\varphi_0}{\eta}\overline{E}. \tag{6.130}$$

This factor f accounts for the variation in the local electric field throughout the EDL. The total velocity of the ions or the particles is the vector sum of both particle motion and bulk fluid motion. Therefore, the total velocity of the ions is given by

$$\overline{u_i} = \overline{u} + \overline{u_{EP,i}}. \tag{6.131}$$

Note that the bulk motion could be due to the external pumping power.

In the case of electro-osmosis, an exponential decay of the concentration from the wall is assumed because there is no fluid motion. However, in this case, as the particles are migrating, they have both advection and diffusion components. Hence, both components should be solved together.

Therefore, in the absence of chemical reactions, the mechanisms of diffusion and convection lead to a flow of species.

Diffusion

$$\bar{j}_{\text{diff},i} = -D_i \nabla c_i, \tag{6.132}$$

where j_{diff} is the diffusive flux density of the species, D_i is the diffusivity of species, i is the solvent, and c_i is the concentration of the species.

Convection

$$\bar{j}_{\text{conv},i} = \overline{u_i} c_i, \tag{6.133}$$

where $\bar{j}_{\text{conv},i}$ is the convective flux density and u_i is the velocity of the species.

Thus, the transport of the species i in the absence of chemical reactions can be described by the Nernst–Planck equation:

$$\frac{\partial c_i}{\partial t} = -\nabla.\left(-D_i \nabla c_i + \overline{u_i} c_i\right). \tag{6.134}$$

This is a well-known equation that does not have any chemical reaction. Apart from the electrophoretic velocities, the Nernst–Planck equation should be solved to observe the distribution of ions, particles, or chemical species in a bulk fluid.

In this chapter, gas and liquid flow, and heat transfer in microchannels are explained clearly. Various experimental correlations in various regimes for both gas and liquid flows are given. Advanced topics such as electrokinetics to solve the N–S equations at small scales are illustrated clearly.

Exercise Problems

1. A flow of helium is generated in a circular microtube by axial pressure and temperature gradients. The microtube has a length $l = 5\,\text{mm}$ and uniform cross section with a diameter $2r_2 = 10\,\mu\text{m}$. In a section where the flow may be considered as locally fully developed, the gradients are $dp/dz = -10\,\text{Pa}/\mu\text{m}$ and $dT/dz = 0.5\,\text{K}/\mu\text{m}$, for a pressure $p = 100\,\text{kPa}$ and a temperature $T = 400\,\text{K}$. Assume a tangential momentum accommodation coefficient $= 0.9$.

a. Write the velocity distribution as a function of the Knudsen number in a dimensionless form: $u_z * (r*, Kn)$.
b. Calculate the part of the velocity increase due to slip at the wall.
c. Calculate the part of the velocity due to thermal creep.

2. Microchannels are directly etched into silicon in order to dissipate 170 W from a computer chip over an active surface area of 10 mm × 10 mm. The geometry may be assumed similar to given figure. Each of the parallel microchannels has a width $a = 40$ m, depth $b = 350$ m and a spacing $s = 30$ m. The silicon thermal conductivity may be assumed to be $k = 180$ W/mK.
   Assume a uniform heat load over the chip base surface and a water inlet temperature of 35 °C. Assume one-dimensional steady-state conduction in the chip substrate.

   a. Calculate the number of flow channels available for cooling.
   b. Assuming the temperature rise of the water to be limited to 10 °C, calculate the required mass flow rate of the water.
   c. Calculate the flow Reynolds number using the mean water temperature for fluid properties.
   d. Check whether the fully developed flow assumption is valid.
   e. Calculate the average heat transfer coefficient in the channels.
   f. Calculate the fin efficiency.
   g. Assuming the heat transfer coefficient to be uniform over the microchannel surface, calculate the surface temperature at the base of the fin at the fluid inlet and fluid outlet sections.
   h. Calculate the pressure drop in the core of the microchannel.
   i. If two large reservoirs are used as the inlet and outlet manifolds, calculate the total pressure drop between the inlet and the outlet manifolds.

3. A microchannel is etched in silicon. The microchannel surface is intentionally etched to provide an average roughness of 12 μm. The microchannel dimensions measured from the root of the roughness elements are width $= 100$ m, height $= 200$ μm, and length $= 10$ mm. Water flows through the microchannels at a temperature of 300 K. Calculate the core frictional pressure drop when $\dot{m} = 90 \times 10^{-6}$ Kg/s, 180×10^{-6} Kg/s.

4. Consider a system in which an electric field is applied across an open rectangular capillary with depth $2d$, width w, length L, and cross-sectional area $A = wd$. Assume that $r \ll w \ll L$ and no pressure gradient are applied. If the wall electro-osmotic mobility is given, derive the pressure per voltage achievable by the pump.

5. Consider a microdevice consisting of two infinite parallel plates separated by a distance $2h$ and filled with an aqueous solution. Assume that the top plate is glass and has an electro-osmotic mobility of 4×10^{-8} m^2/Vs and that the bottom plate is Teflon and has an electro-osmotic mobility of 2×10^{-8} m^2/Vs. Derive the resulting velocity distribution if an electric field of 150 V/cm is applied.

6. Calculate λD for the following conditions. In each case, assume that the system is well approximated by a symmetric electrolyte with the valence "z" and concentration "c" specified in parentheses.

a. An electrolyte solution with $z = 1$, $c = 1 \times 10^{-7}$ M. This might represent deionized water.

b. An electrolyte solution with $z = 1$, $c = 2 \times 10^{-6}$ M. This might represent water at equilibrium with air—in this case, there is dissolved CO_2 and equilibrium HCO_3^-.

c. An electrolyte solution with $z = 2$, $c = 1 \times 10^{-3}$ M. This might represent 1-mM Epsom salts.

d. An electrolyte solution with $z = 1$, $c = 0.18$ M. This might approximate phosphate-buffered saline (PBS). This is often used for biochemical analysis, wound cleaning, or temporary cell storage.

Chapter 7
Phase Change in Minichannels and Microchannels

7.1 Boiling on Thin Devices

Phase change involves two phases in a single component, whereas general two-phase flows can be adiabatic with multiple phases. One phase can be air and the other can be water. Two-phase flows are used in a wide range of industrial applications such as atmospheric sciences, biological sciences, and heat transfer processes. In heat transfer, the main focus is on phase change that involves two distinct phases in a single component. Therefore, typically this kind of phenomenon is applied in electronic cooling, where microchannels and minichannels are used to increase the heat transfer rate. The fundamental advantage of two-phase flow boiling is that the two-phase heat transfer coefficient is three to four orders of magnitude higher than the single-phase heat transfer coefficient.

Professor Nukiyama has conducted experiments on boiling (from which we can obtain greater insights into the boiling phenomenon) and invented the boiling curve of heat flux vs wall superheat $(T_w - T_{sat})$, also called the Nukiyama boiling curve, where T_w and T_{sat} are the wall temperature and the saturation temperature, respectively. Nukiyama submerged a heater coil in static water and connected the system to a DC power supply. The power input was varied to control the heat flux such that the heat flux value gradually increases. Thermocouples were placed on the surface of the heater coil to measure the wall temperature of the heater coil.

In the initial stage, as the heat flux increases, the degree of wall superheat $(T_w - T_{sat})$ also increases and bubbles are not formed. This is the point where the wall temperature is just above the saturation temperature. The saturation temperature of water at 1 atm is approximately 100 °C. On maintaining the wall temperature slightly above the saturation temperature (100.5 °C), the liquid layer which is in contact with the heater is close to the saturation temperature, whereas the bulk liquid is approximately at room temperature. Therefore, under these conditions, the mode of heat transfer is purely natural convection. Hence, natural convection currents drive the heat from the fluid surrounding the heater to the surface. The slope of this curve

A. Pattamatta and S. K. Das, *Fundamentals of Nano- and Microscale Heat Transport*,
https://doi.org/10.1007/978-3-031-89613-2_7

gives the heat transfer coefficient, i.e.,

$$h = \frac{q_w}{T_w - T_{sat}}.$$

(7.1)

This phenomenon continues up to a certain heat flux value. On increasing the heat flux beyond this value, the wall temperature increases significantly. At this stage, the degree of superheat could be of the order of $100\,^\circ$C. Therefore, the liquid near the heater surface is reasonably well above the saturation temperature. The boiling phenomenon commences at this stage and the liquid slowly starts to evaporate. Evaporation is a surface phenomenon driven by mass transfer.

Consider a system with a water reservoir and ambient air with a relative humidity of 10%. Therefore, very little water vapor is saturated in the ambient air. The partial pressure of water vapor in the air is very low compared to the overall pressure.

Therefore, the difference between the partial pressure at the interface and the partial pressure of ambient air drives evaporation. A few liquid molecules excited at the interface have enough kinetic energy and become vapor due to the very high potential gradient outside. The conversion of liquid molecules to the vapor state continues further until there is an increase in the relative humidity.

If the relative humidity reaches 100%, then the evaporation process ends. Therefore, evaporation is purely a mass transfer phenomenon and is driven by the concentration gradient. This can happen at any temperature and not necessarily above the saturation temperature, whereas boiling is a bulk phenomenon, which is not specifically a surface process. Boiling can happen throughout the entire bulk of the liquid and is driven by temperature gradients.

The phase-change phenomenon is a process too complicated to imagine. Therefore, a typical explanation is that the heater itself should have some roughness elements that are rough enough to induce artificial cavitation sites for bubble growth.

At the nanoscale, cavities are formed by the roughness elements on the heater, which hold the water vapor. For the phase-change phenomenon to occur, the vapor must grow to a sufficient level so that the pressure at the cavities can be higher than the surrounding liquid pressure. Therefore, fine clerical bubbles can be formed, and these bubbles should be able to rise all the way because of buoyancy, and at the same time the pressure difference should be high enough for the bubble to sustain till the top of the surface. They should have much larger diameters near the surface so that they can rise continuously till the surface, and finally, these bubbles will collapse at the surface.

In the case of very low superheats, bubbles will form near the heater surface, but they may not have enough strength to rise all the way up to the top. Therefore, they will just disappear. However, on increasing the superheat, these cavities can hold high volumes of vapor with sufficient vapor pressure. These bubbles grow because of a higher vapor pressure as the vapor pressure increases with temperature. If the pressure is more, then the bubbles can rise to various heights, and they will collapse eventually at the interface.

This is a continuous process. On increasing the heat flux, the slope of the boiling curve also increases. In the natural convection regime, very moderate values of the heat transfer coefficient can be observed. The phenomenon of bubble nucleation and bubble growth starts with increasing heat flux. This regime is called the nucleate boiling regime. In this regime, as superheat increases, both the size and number of bubbles increase.

After a certain amount of time, on increasing the amount of superheat, complete conversion of the liquid to vapor takes place. The vapor film that coats the entire heater surface causes a reduction in heat transfer coefficient because the vapor has lower thermal conductivity compared to water. Therefore, the slope of the boiling curve diminishes. This is called the film boiling regime. In practical cases, the curve will not go down because of the continuous addition of heat flux.

In this film boiling regime, the temperature of the heater surface increases tremendously, but the heat transfer rate is not sufficient for the heat to dissipate directly from the heater surface to the bulk. After a particular time, the surface temperature reaches the melting point of the metal and the entire heater breaks immediately.

The maximum value of the heat flux above which the surface temperature of the heater increases significantly is called the critical heat flux (CHF). Therefore, it is not advisable to operate the heat flux beyond CHF. This CHF value depends on the type of fluid and surface.

On controlling the heat flux appropriately, one cannot obtain a downfall region in the boiling curve. To do this, the surface temperature must be controlled rather than heat flux. Therefore, a heat exchanger must be incorporated and the heater should be maintained under isothermal conditions, and the temperature of the heater should be increased slowly.

Bubbles can be observed in the case of flow through a channel or a duct with a constant wall flux or at a constant wall temperature. These bubbles move along with the bulk fluid and the vapor volume fraction increases as one moves from left to right. Therefore, depending on the vapor volume fraction, a two-phase mixture with different regimes can be observed. From the boiling curve, in nucleate boiling regime, one can note that the two-phase heat transfer coefficient is three orders of magnitude higher than that of the single-phase heat transfer coefficient. Therefore, it is always beneficial to operate two-phase flows in either macrochannels or microchannels.

The most significant difference between the macrochannels and microchannels is the effect of gravity. In macrochannels, the flow is entirely driven by gravity, whereas in microchannels the surface tension drives the flow. The other significant difference is the relative size of bubbles. In the case of macrochannels, the diameter of the channel is much larger. Therefore, one single bubble cannot occupy the entire channel cross section, whereas in the case of microchannels only one bubble occupies the entire channel cross section and it cannot grow further. Hence, it can be elongated along the length of the channel. Therefore, most of the time, a bubbly regime can be observed in macrochannels where multiple bubbles simultaneously coalesce. Coalescence is the merging of bubbles when they come close to each other. In the case of microchannels, one single nucleating growing bubble quickly fills the channel diameter and then continues to be elongated.

The other significant difference is the channel orientation. The flow regimes change with respect to the orientation. In the case of macrochannels in a horizontal orientation, stratified flow can be observed where the liquid is collected at the bottom and vapor is collected at the top. In the vertical mode of operation, the flow regimes are entirely different. However, the microchannel is entirely independent of orientation. Therefore, it is very advantageous to use microchannels than macrochannels.

7.2   Non-dimensional Numbers Used in Microchannels

Bond Number (Bo)

It is the ratio of the buoyancy force to the surface tension force. It is widely used in droplet and spray applications. Microchannels can be classified based on the Bond number; for classification, the Bond number should be less than 2.5 so that the surface tension force is more dominant than the gravitational force and is given as

$$Bo = \frac{g\,(\rho_L - \rho_v)\,D^2}{\sigma}.$$ (7.2)

Eotvos Number (Eo)

The Bond number can also be called Eotvos number. The Eotvos number is the square root of the Bond number. It can be used at low velocities and vapor fractions, and is given as

$$Eo = \frac{g\,(\rho_L - \rho_v)\,L^2}{\sigma}.$$ (7.3)

Capillary Number (Ca)

The capillary number is the ratio of viscous force to surface tension force. Therefore, it is significant in flow-through microchannels, where the order of surface tension force to the inertia or viscous forces has to be classified. It is also used in bubble removal analysis. The capillary number is expressed as follows:

$$Ca = \frac{\mu V}{\sigma}.$$ (7.4)

Ohnesorge Number (Z)

It is the ratio of viscous force to the square root of inertia and surface tension force. It is widely used in automization studies and is less commonly used in microchannels. It is expressed as follows:

$$Z = \frac{\mu}{(\rho L \sigma)^{1/2}}.$$

(7.5)

Weber Number (We)

It is the ratio of the inertia force to the surface tension force. It is used to study the relative effects of surface tension and inertia forces on flow patterns in microchannels. It is expressed as follows:

$$We = \frac{LG^2}{\rho \sigma}.$$

(7.6)

Jakob Number (Ja)

It is the ratio of the sensible heat required for reaching a saturation temperature to the latent heat. The aforementioned non-dimensional numbers describe the hydrodynamic characteristics, whereas the Jakob number characterizes heat transfer.

If the sensible heat is greater than the latent heat, the operating regime is in a single phase. In this case, the Jakob number is higher, and it is low if the latent heat is greater than the sensible heat. Therefore, significant values of the Jakob number indicate the single-phase regime and small values of the Jakob number indicate the phase change regime. The expression for the Jakob number is given as

$$Ja = \frac{\rho_L}{\rho_v} \frac{c_{p,L} \Delta T}{h_{LV}}.$$

(7.7)

Martinelli Parameter (X)

In two-phase flows, there exists a mixture of liquid and vapor. Therefore, to calculate the two-phase pressure drop, the Martinelli parameter can be used. The Martinelli parameter is defined as the relative ratio of the pressure drop of liquid to the pressure drop of vapor flowing alone and is expressed as follows:

$$X^2 = \left(\left. \frac{dp}{dz} \right|_F \right)_L \bigg/ \left(\left. \frac{dp}{dz} \right|_F \right)_V.$$

(7.8)

Generally, the liquid pressure drop is higher than the vapor pressure drop. Therefore, this ratio is greater than unity, and it is used to develop correlations to predict pressure drop for two-phase mixtures. These correlations are the functions of the Martinelli parameter because this quantity can be easily found as the friction factor depends on the standard Moody diagram. Hence, one can find the pure single-phase liquid or pure single-phase vapor pressure drop. The Martinelli parameter is the ratio of these two quantities. The Martinelli parameter is used in two-phase pressure drop models. Therefore, one can build the correlations for the actual two-phase case as a function of two single-phase pressure drops.

Convection Number (Co)

The convection number can be defined as

$$C_0 = \left[\frac{1-x}{x}\right]^{0.8} \left(\frac{\rho_v}{\rho_L}\right)^{0.5}, \tag{7.9}$$

where x is the mass fraction and it can be defined as the ratio of the mass of the vapor to the sum of the mass of the liquid and mass of the vapor. This number signifies the type of operating regime such as pure liquid, pure vapor, and a mixture of liquid and vapor. Based on the type of regime, the correlations can be built for heat transfer rate and heat transfer coefficient as a function of convection number.

Boiling Number (Bo)

The boiling number signifies the relative importance of heat-flux-driven evaporation to mass-flux-driven evaporation, i.e.,

$$Bo = \frac{q''}{Gh_{LV}}, \tag{7.10}$$

where q'' is the heat flux, G is the mass flux, and h_{LV} is the latent heat of vaporization. Usually, this parameter can only be used in flow boiling and not in pool boiling.

K_1 It is the ratio of evaporation momentum force to inertial force at the liquid–vapor interface. It can be used in flow boiling systems, where surface tension forces become dominant. It is expressed as follows:

$$K_1 = \left(\frac{q''}{Gh_{LV}}\right)^2 \frac{\rho_L}{\rho_V}. \tag{7.11}$$

K_2 It is the ratio of evaporation momentum force to surface tension force at the liquid–vapor interface. It is applicable in modeling interface motion, such as the critical heat flux. It is expressed as follows:

$$K_2 = \left(\frac{q''}{h_{LV}}\right)^2 \frac{D}{\rho_V \sigma}.$$

(7.12)

7.3 Flow Regimes in Macrochannels and Microchannels

7.3.1 Flow Regimes in Macrochannels

The flow regimes in macrochannels are shown in Fig. 7.1. There are two distinct regimes in this figure: the adiabatic two-phase flow and the phase change. Under adiabatic two-phase flow conditions, air and water in different proportions and different mass fractions are mixed without heating. In the case of phase change, various kinds of flow patterns can be observed by heating the tube and passing the liquid into it. In both adiabatic and two-phase cases, flow regimes are studied in both horizontal and vertical modes since microchannels are sensitive to the orientation.

In the case of the vertical orientation, on heating the tube, bubbles will be formed inside it. With a continuous supply of heat, the bubbles elongate and become slugs, and these slugs, elongate further. Therefore, the core is filled with vapor and it is

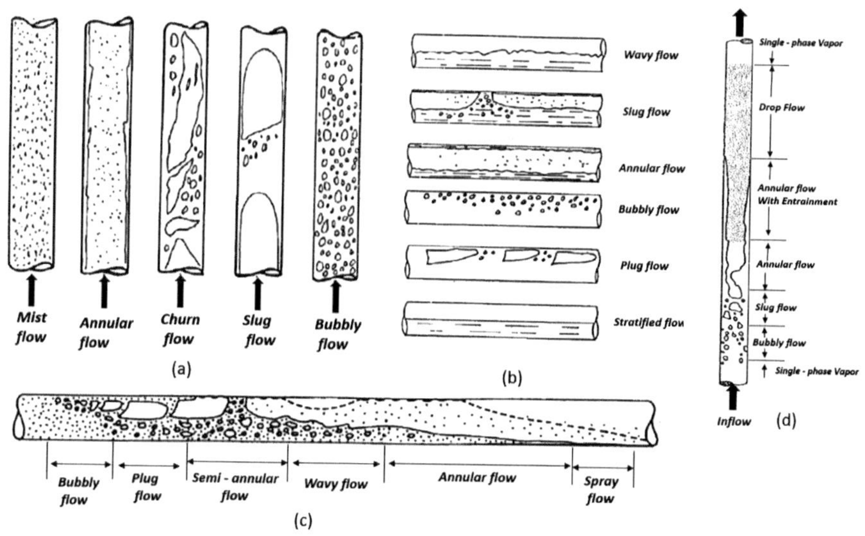

Fig. 7.1 Flow regimes in macrochannels: **a** vertical upward adiabatic; **b** horizontal adiabatic; **c** horizontal boiling; **d** vertical upward boiling

surrounded by a thin liquid film at the annulus, which sticks to the wall. This core increases in size and finally the entire tube diameter cross section is filled with vapor and the liquid is strapped as fine mist within this vapor. This is called the mist flow. This liquid gets entrapped within a complete vapor, and after a specified period, the liquid evaporates and only pure vapor is present inside the tube.

Therefore, as heating progresses, the mass fraction changes from 0 to 1. On continuously heating the tube, all the flow regimes can be observed. In the horizontal orientation, similar flow patterns can be observed except that some patterns are influenced by gravity and the stratification of the liquid can be observed at the bottom of the tube above which the vapor forms.

Therefore, a clear annular flow cannot be observed in the horizontal orientation. In this case, a perfect vapor core is surrounded by the liquid in the horizontal alignment. The liquid settles down at the bottom and the vapor tends to settle above the liquid. However, as the mass fraction increases, a mist flow or a drop flow can be observed, and finally, the entire liquid becomes pure vapor.

7.3.2 Flow Regimes in Microchannels

Bubbly Flow

In microchannels, on applying the heat flux, isolated round and elongated bubbles smaller than the cross section of the microchannels can be formed. This flow regime is termed as bubbly flow. These bubbles move in the flow direction, as shown in Fig. 7.2a. The bubbles generally nucleate at the microchannel walls and detach from the walls after growing. The shape and size of the bubbles vary with the flow rate and the heat flux.

Slug Flow

As the heat flux increases, single nucleating bubbles grow and quickly expand to the diameter of the tube. They cannot grow any further. Therefore, they will only expand along the length of the tube. Hence, in microchannels, an elongated bubbly flow or a slug flow can be observed. This flow is also called a Taylor bubble flow. It can be represented pictorially, as shown in Fig. 7.2b. A fragile film of liquid protects the bubbles directly from the impact of heat transfer from the tube. As the volume fraction of the vapor increases, the slug flow regime is slowly converted into the slug annular regime.

In this regime, most of the vapor fills the core of the tube and the liquid covers the core as an annulus. As the vapor volume fraction increases further, only pure vapor remains in the tube. A well-defined elongated bubble cannot be expected every time

Fig. 7.2 Flow regimes in microchannels: **a** bubbly flow; **b** slug flow; **c** churn flow; **d** wispy-annular flow; **e** annular flow

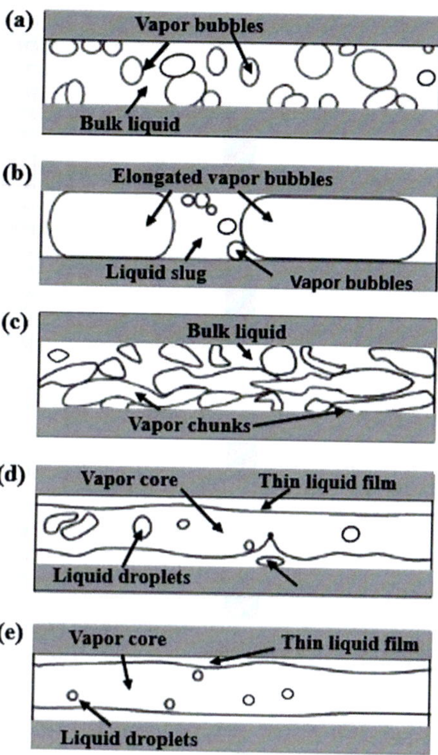

in the case of slug flow. Sometimes, hysteresis can be observed in the advancing and receding meniscuses of these Taylor bubbles.

The shape of the meniscus may be different for the advancing side, which is moving probably in the direction opposite to the motion of the liquid in the receding side. The advancing and receding sides are defined with respect to the direction of the liquid flow. Therefore, the advancing and receding meniscuses have different shapes based on the temperature along the length of the tube. On applying a uniform heat flux, the wall temperature increases continuously along the length of the tube. This causes the hysteresis in the shape of the meniscus in the advancing and receding sections. Various kinds of slug flows, such as churn flows and wispy-annular flows, are rarely encountered in microchannels.

Churn Flow

In a churn flow, small bubbles exist in the liquid slugs between the elongated bubbles, as shown in Fig. 7.2c. This flow regime consists of vapor churns transported downstream and large bubbles nucleating at a high rate at the channel walls. At high heat fluxes, nucleation at the walls may be suppressed.

Wispy-annular Flow

In a wispy-annular flow, a vapor core is separated from the channel walls by a relatively thick and unstable liquid film, as shown in Fig. 7.2d. Large, irregular-shaped droplets are entrained in the vapor core. Very few nucleation sites remain in the liquid film, which results in small vapor bubbles in the liquid layer.

Annular Flow

In an annular flow, the liquid layer is thinner than that in a wispy-annular flow. The liquid film thickness decreases as the heat flux increases. Small, round droplets are entrained in the vapor core, while no vapor bubbles are seen in the liquid annulus, as shown in Fig. 7.2e. At very high heat fluxes, when critical heat flux is reached, the walls can completely dry out. Under certain conditions, a vapor blanket forms at the walls around a liquid core, which flows through the center of the channels. Therefore, on increasing the heat flux continuously, various flow regimes such as bubbly flow, slug flow, and annular flow occur successively in microchannels.

The dry-out phase will not promote heat transfer. Therefore, most of the flow regimes in microchannels are confined toward the slug flow.

7.3.3 Flow Maps

As the name suggests, they map different flow regimes in a 2-D diagram in which the mass flow rates for vapor and liquid are plotted on the X-and Y-axes, respectively.

If the vapor content is more than liquid content, then there is a slip between the liquid and the vapor and both will not have the same velocity. There is a possibility of obtaining a higher liquid velocity compared to gas and vice versa. Therefore, depending on the relative magnitude of one over the other, all these flow regimes can be drawn on a 2-D diagram and marked as flow maps.

Figure 7.3 shows the flow maps for a typical tube diameter of 3.4 mm. If the values of velocity for both the liquid and vapor are very low, then the type of operating regime is bubbly flow. The slug flow can be observed at very low liquid and moderate gas flow rates. On increasing the gas flow rate further, an annular regime can be observed. In the case of minitubes, at $D = 1.2$ mm, the size of the slug flow regime increases and becomes dominant for a limited range of gas velocities, as shown in Fig. 7.4. In

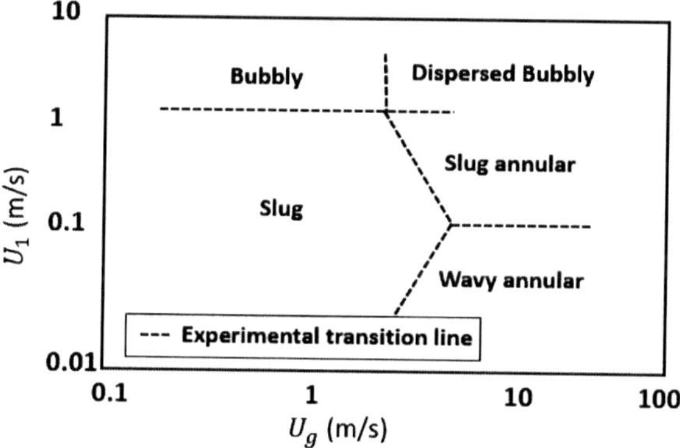

Fig. 7.3   Flow maps for a typical tube diameter of 3.4 mm

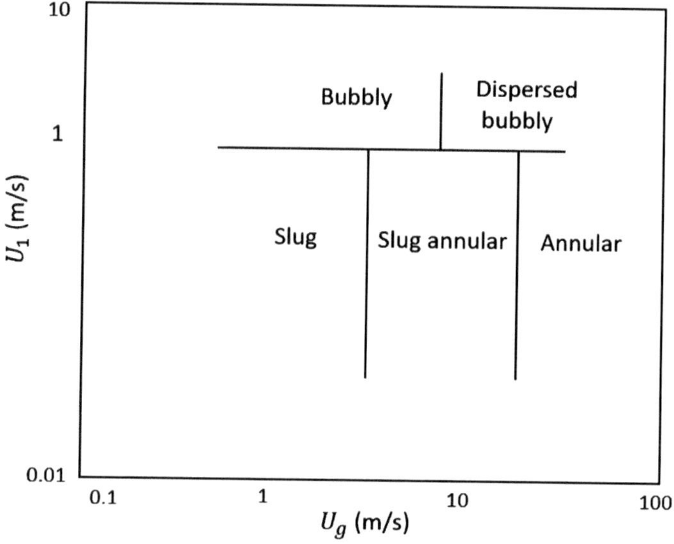

Fig. 7.4   Flow map for a typical tube diameter of 1.2 m

microchannels, at $D = 0.6$ mm, slug flow can be observed for a wide range of gas velocities, as shown in Fig. 7.5.

Finally, the pattern becomes a slug or a slug–annular. Therefore, in microchannels, the flow pattern depends only on liquid velocity, irrespective of gas velocity. If the liquid velocity is moderately high, then a slug flow regime can be observed, and if it drops, the flow regime becomes annular. Therefore, only two regimes, slug and annular, can be observed. In microchannels, irrespective of the velocity of the gas, by controlling the velocity of the liquid, a slug flow can be maintained. Therefore, a slug flow is the most predominant flow pattern in microchannels.

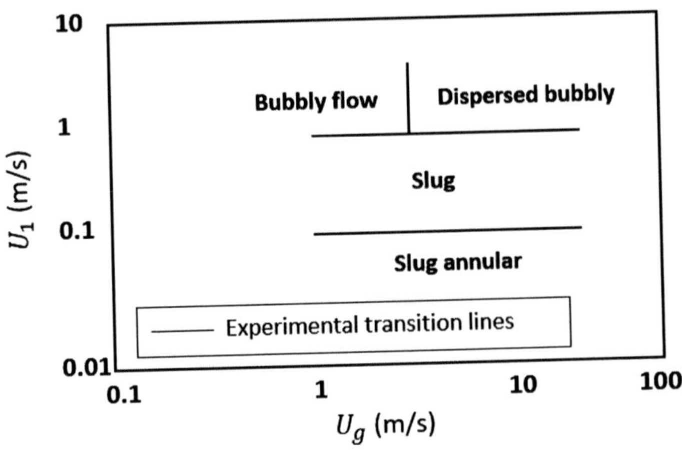

Fig. 7.5 Flow map for a typical tube diameter of 0.61 mm

7.3.4 Pressure Drop Estimation in Microchannels

Pressure drop estimation is a critical parameter in microchannels. In microchannels, as the tube diameter is very small, a large pressure drop penalty can be observed. As the diameter of the tube is very small, for the same value of friction factor, a large pressure drop is required to drive the flow for the same Reynolds number. Therefore, pressure drop estimation becomes very important.

The pressure drop in a microchannel or a minichannel heat exchanger is the sum of the following components:

$$\Delta p = \Delta p_c + \Delta p_{f,1-ph} + \Delta p_{f,tp} + \Delta p_a + \Delta p_g + \Delta p_e, \tag{7.13}$$

where Δp_c is the pressure drop due to contraction at the entrance, $\Delta p_{f,1-ph}$ is the single-phase pressure loss due to friction, including entrance region effects, $\Delta p_{f,tp}$ is the two-phase frictional pressure drop, Δp_a is the pressure drop due to acceleration associated with evaporation, Δp_g is the pressure drop due to gravitational force, and Δp_e is the pressure drop due to expansion at the outlet.

The influence of gravitational head can be neglected. The calculation of single-phase pressure loss due to friction is taken from the Darcy chart by filling the entire tube with liquid. Δp_c is the pressure loss due to sudden contraction and is given as follows:

$$\Delta p_c = \frac{G^2}{2\rho_L}\left[\left(\frac{1}{C_o} - 1\right)^2 + 1 - \frac{1}{\sigma_c^2}\right]\psi_h, \tag{7.14}$$

where G is the mass flux, σ_c is the contraction area ratio (header to channel > 1) and C_0 is the contraction coefficient, which is given as

$$C_0 = \frac{1}{0.639\left(1 - \frac{1}{\sigma_c}\right)^{0.5} + 1}. \tag{7.15}$$

ψ_h is the two-phase homogeneous flow multiplier and is given as

$$\psi_h = \left[1 + x\left(\rho_L/\rho_V - 1\right)\right], \tag{7.16}$$

where x is the quality. The exit pressure loss can be calculated as follows:

$$\Delta p_e = G^2 \sigma_e \left(1 - \sigma_e\right) \psi_s, \tag{7.17}$$

where σ_c is the area expansion ratio and ψ_s is the separated flow multiplier and is expressed as

$$\psi_s = 1 + \left(\frac{\rho_L}{\rho_V} - 1\right)\left[0.25x\left(1 - x\right) + x^2\right]. \tag{7.18}$$

The two-phase frictional pressure drop can be obtained by multiplying the pressure drop with a two-phase multiplier when the tube is filled with a pure liquid. The two-phase multiplier is the ratio of the pressure drop when the tube is filled with liquid to the pressure drop when it is filled with gas. Therefore, the local friction pressure gradient at any section for a two-phase flow can be calculated by

$$\left(\frac{dp_F}{dz}\right) = \left(\frac{dp_F}{dz}\right)_L \phi_L^2, \tag{7.19}$$

where ϕ_L^2 is the two-phase multiplier and is given by

$$\phi_L^2 = 1 + \frac{C}{X} + \frac{1}{X^2}, \tag{7.20}$$

where X is the Martinelli parameter.

The constant C is known as the Chisholm parameter and depends on the type of flow regime as follows:

- $C = 21$ if both phases are turbulent,
- $C = 12$ for laminar liquid and turbulent vapor,
- $C = 10$ for turbulent liquid and laminar vapor,
- $C = 5$ if both phases are laminar.

Mishima and Hibiki [75] obtained the equation for C as follows:

$$C = 21 \left(1 - e^{-319D_h}\right). \tag{7.21}$$

The acceleration pressure drop can be calculated as

$$\Delta p_a = G^2 v_{LV} x_e, \tag{7.22}$$

where x_e is the exit quality. v_{LV} is the difference between the specific volumes of vapor and liquid phases.

$$v_{LV} = v_V - v_L. \tag{7.23}$$

The gravitational pressure drop can be expressed as

$$\Delta p_g = \frac{g \, (\sin \beta) \, L}{v_{LV} x_e} \ln \left(1 + x_e \frac{v_{LV}}{v_L}\right). \tag{7.24}$$

As the diameter of the channel decreases, the gravitational pressure drop becomes negligible. Therefore, the two-phase pressure drop due to friction is the biggest contributor to pressure drop in microchannels. Also, the pressure drops due to both entry and exit losses are significant because of very high contraction and expansion ratios.

Figure 7.6 shows the variation of pressure drop with mass quality. Some deviations can be observed by comparing the overall pressure drop from the correlations with experimental data, for example, while using the value of C from Mishima and Hibiki, there is a substantial deviation from the experimental data. The other model for C proposed by English and Kandlikar [37] shows good agreement.

Fig. 7.6 Variation of pressure drop with mass quality

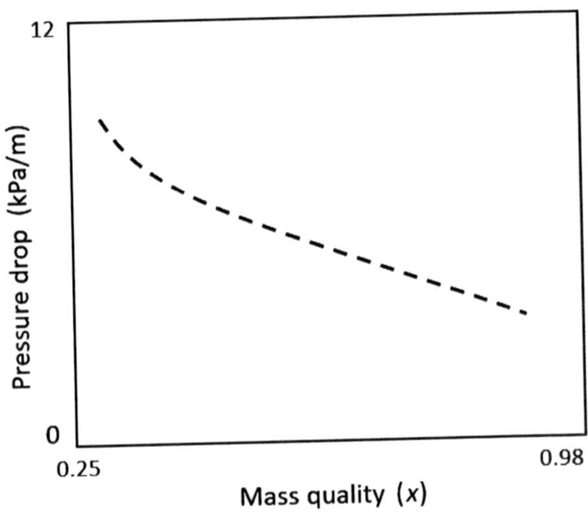

7.3.5 *Heat Transfer Correlations in Microchannels*

There are no analytical methods to calculate the heat transfer coefficient in two-phase flows. Kandlikar and Steinke [62] and Kandlikar and Balasubramanian [59] introduced the laminar flow equations as follows.

When the flow becomes turbulent, from these correlations, the single-phase heat transfer coefficient is calculated and substituted into the expressions for the two-phase heat transfer coefficient ($h_{TP,NBD}$ and $h_{TP,CBD}$). The larger value of $h_{TP,NBD}$ and $h_{TP,CBD}$ gives the two-phase heat transfer coefficient. For $Re_{LO} > 100$ (LO — liquid only), h_{TP} is the larger value of $h_{TP,NBD}$ and $h_{TP,CBD}$.

$$h_{TP,NBD} = 0.6683 Co^{-0.2}(1-x)^{0.8}h_{LO} + 1058.0 Bo^{0.7}(1-x)^{0.8}F_{Fl}h_{LO},$$
(7.25)

$$h_{TP,CBD} = 1.136 Co^{-0.9}(1-x)^{0.8}h_{LO} + 667.2 Bo^{0.7}(1-x)^{0.8}F_{Fl}h_{LO},$$
(7.26)

where h_{LO} is the single-phase heat transfer coefficient, which can be calculated by filling the entire tube with liquid. The convection number (Co) is given by

$$Co = [(1-x)/x]^{0.8}(\rho_v/\rho_L)^{0.5}.$$
(7.27)

The boiling number (Bo) is given by

$$Bo = q''/Gh_{LV}.$$
(7.28)

We know that x is the quality of vapor mass fraction and F_{Fl} is the fluid surface parameter. The values of F_{Fl} for various fluids are tabulated in Fig. 7.7. The single-phase liquid flow heat transfer coefficient h_{LO} is given as follows:

For $10^4 \leq Re_{LO} \leq 5 \times 10^6$,

$$h_{LO} = \frac{Re_{LO}Pr_L\,(f/2)\,(k_L/D)}{1 + 12.7\left(Pr_L^{2/3} - 1\right)(f/2)^{0.5}}.$$
(7.29)

Fig. 7.7 Typical F_{Fl} (fluid surface parameter) values in the flow boiling correlations

Fluid	F_{F1}
Water	1
R-11	1.30
R-12	1.50
R-22	2.20
R-134a	1.63

For $3000 \leq \mathrm{Re}_{LO} \leq 10^4$,

$$h_{LO} = \frac{(\mathrm{Re}_{LO} - 1000)\,\mathrm{Pr}_L\,(f/2)\,(k_L/D)}{1 + 12.7\left(\mathrm{Pr}_L^{2/3} - 1\right)(f/2)^{0.5}}. \tag{7.30}$$

For $100 \leq \mathrm{Re}_{LO} \leq 1600$,

$$h_{LO} = \frac{Nu_{LO}k}{D_h}. \tag{7.31}$$

For very small Reynolds numbers, the correlation for heat transfer coefficient is purely based on $h_{TP,NBD}$. Therefore, the value of $h_{TP,CBD}$ need not be calculated. For $Re < 100$, the nucleate boiling regime governs, and the correlation becomes

$$h_{TP} = h_{TP,NBD} = 0.6683Co^{-0.2}(1-x)^{0.8}h_{LO} + 1058.0Bo^{0.7}(1-x)^{0.8}F_{Fl}h_{LO}, \tag{7.32}$$

where NBD and CBD represent nucleate boiling and convective boiling, respectively. For high Reynolds numbers, there are contributions from both nucleate boiling and convective boiling, whereas for low Reynolds numbers, the contribution of convective boiling can be ignored.

These correlations show that these are functions of vapor volume fraction or mass fraction x. Heat transfer coefficient values are plotted for different values of quality and heat flux by heating the tube continuously in vertical orientation, as shown in Fig. 7.8. These results are compared to the experimental results of Kandlikar [57], and both are in perfect conformity. The figure shows that a two-phase heat transfer coefficient reaches a maximum at $x = 0.2$, which corresponds to the slug flow regime. As the mass fraction increases, the flow regime changes into annular flow and finally dry-out occurs. Therefore, the two-phase heat transfer coefficient drops from slug flow toward dry-out. The difference between nucleate boiling and convective boiling

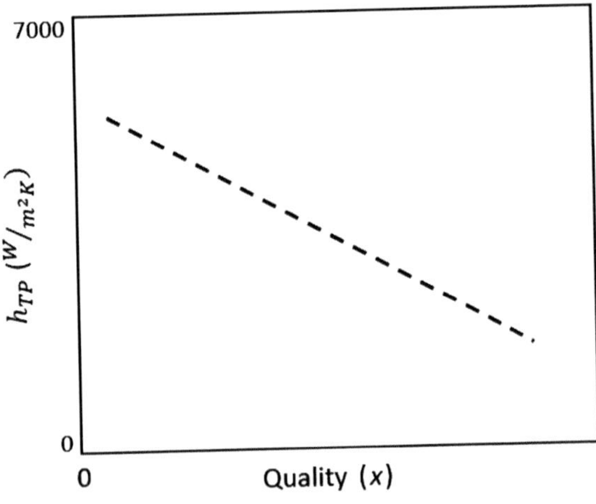

Fig. 7.8 Variation of heat transfer coefficient with quality

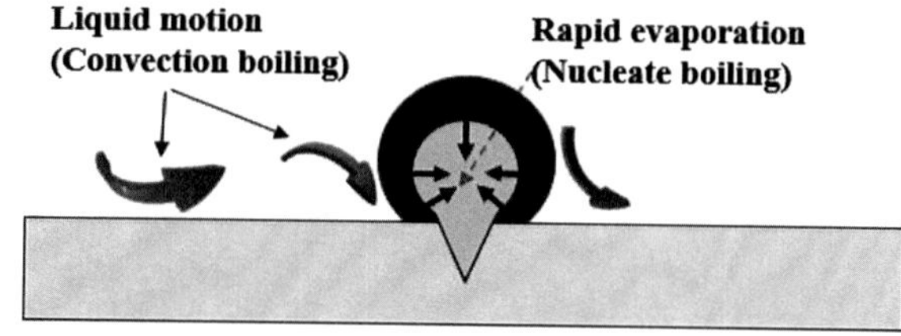

Fig. 7.9 Convective and nucleate boiling phenomena

is that, in nucleate boiling, the initial nucleation and evaporation occur due to the presence of nucleation sites.

If the local temperature exceeds the saturation temperature, the initiation and growth of bubbles start and the bulk of the liquid starts to boil. In convective boiling, the augmentation of evaporation rates can be increased by inducing the bulk motion in the liquid by stirring. A pictorial representation of nucleate and convective boiling regimes is shown in Fig. 7.9.

Therefore, the overall effect of flow boiling is due to stirring and localized evaporation due to temperature gradient. This is why the heat transfer coefficient is higher in the flow boiling regime because convective boiling contributes to nucleate boiling.

The two-phase heat transfer phenomenon in microchannels is explained in this chapter. Various flow regimes, non-dimensional numbers used in microchannels, and heat transfer correlations are illustrated in detail.

Exercise Problems

1. Water is used as the cooling liquid in a microchannel heat sink. The dimensions of one channel are $a = 1054\,\mu m \times b = 50\,\mu m$, where "$a$" is the unheated length in the three-sided heating case. The inlet temperature of the water is $70\,°C$, and the Reynolds number in the channel is 600. Quality of the steam at exit of (i) $q'' = 50\,kW/m^2$ is 0.01 and that of (ii) $q'' = 340\,kW/m^2$ is 0.055. Initial 10 mm length has single-phase liquid followed by phase change for remaining length.

   (a) Plot the predicted heat transfer coefficient as a function of quality for the following values of heat flux:

   (i) $q'' = 50\,kW/m^2$, (ii) $q'' = 340\,kW/m^2$.

   (b) Calculate the pressure drop in the heat exchanger core for a heat flux of (i) $q'' = 50\,kW/m^2$ (ii) $q'' = 340\,kW/m^2$ a channel length of 20 mm.

Assumptions: Fully developed laminar flow Nusselt number; three-sided heating; properties of water are at saturation temperature at 1 atm; Receding contact angle is 40°.

Properties: Properties of water at $100\,^\circ$C:

1. $h_{LV} = 2.26 \times 10^6$ J/kg,
2. $i_L = 4.19 \times 10^5$ J/kg,
3. $\rho_V = 0.596$ kg/m^3,
4. $v_L = 0.001044$ m^3/kg,
5. $v_g = 1.679$ m^3/kg,
6. $c_{p,L} = 4217$ J/kgK,

7. $\mu_V = 1.20 \times 10^{-5}$ Ns/m^2,
8. $\mu_L = 2.79 \times 10^{-4}$ Ns/m^2,
9. $k_L = 0.68$ W/mK,
10. $\sigma = 0.0589$ N/m,
11. $Pr = 1.76$

(Incropera and DeWitt [54]).

Properties of water at $70\,^\circ$C:

1. $i_L = 2.93 \times 10^5$ J/kg.

Chapter 8
Nanofluids

8.1 Introduction to Nanofluids

Nanofluids are gaining importance nowadays in cooling applications due to their enhanced thermophysical and heat transfer characteristics. As researchers treat nanofluids as bulk macroscale continuum fluids, it is challenging to understand the nanoscale aspects of a nanofluid. From the heat transfer point of view, nanofluids can be treated as a bulk fluid with a separate set of properties. Nanofluids may be defined as a dilute colloidal suspension of nanometer-sized particles dispersed in a base fluid, as shown in Fig. 8.1. A colloidal dispersion is formed by mixing the insoluble solid particles with base fluids, such as water or ethylene glycol. The examples of nanofluids include Al_2O_3–water, CuO–water, and Cu–water.

Producing a stable suspension of nanoparticles is one of the challenges faced during the preparation of nanofluids. Since the density of nanoparticles is much higher than that of base fluids, gravitational force acts on the nanoparticles. The only problem with nanofluids is the increase in pressure drop because of the generous size of the particles. Problems such as fouling, clogging, sedimentation, and corrosion occur due to conventional slurries because these particles are very large; as a result, they can damage coolant tubes.

Reduction of the size of these particles does not only lead to minimal increase of pumping power over the base fluid, but other problems such as erosion and corrosion can be minimized. Also, it is easier to suspend these smaller nanoparticles in the base fluid because of the higher surface area-to-volume ratio. Stronger surface tension forces lead to the agglomeration of nanoparticles. Since the higher surface tension force tries to push these particles, the clogging or agglomeration of particles can be avoided.

In nanofluids, the electric double-layer formation becomes very dominant because the surface area-to-volume ratio is higher. Electrostatic forces keep the particles stable, thus preventing the agglomeration of particles. Conventional cooling fluids

© The Author(s) 2025

A. Pattamatta and S. K. Das, *Fundamentals of Nano- and Microscale Heat Transport*,
https://doi.org/10.1007/978-3-031-89613-2_8

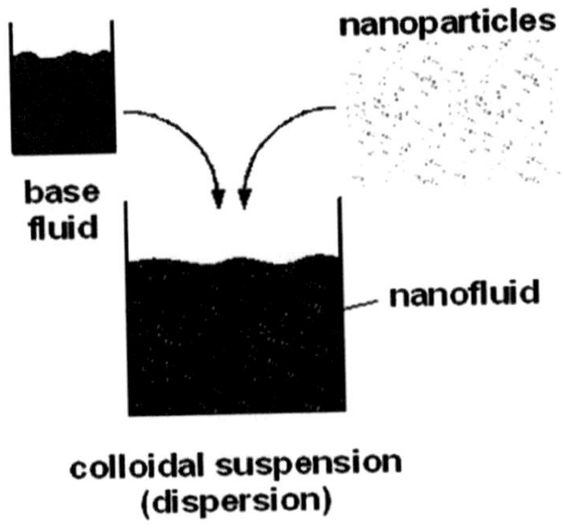

Fig. 8.1 Schematic representation of a nanofluid

can be replaced with nanofluids because of the advantages due to their thermophysical and heat transfer characteristics.

Nanofluids can be used as coolants, lubricants, and also in drug delivery systems. The cooling applications of nanofluids include electronic cooling, engine cooling, and transformer cooling. Zhang et al. (1997) reported that the surface-modified nanoparticles dispersed in mineral oils could effectively reduce the wear-enhancing load-carrying capacity. They can also be used for biomedical applications, for example, the selective destruction of tumors using a laser.

8.1.1 Preparation of Nanofluids

Nanofluids can be prepared by two methods: a one-step process and a two-step process. In the one-step process, nanoparticles can be synthesized directly and dispersed into base fluids simultaneously to form nanoparticle suspensions. The Hammer method is used for the chemical synthesis of nanofluids, for example, graphene nanofluids.

In the two-step process shown in Fig. 8.2, nanoparticles are prepared mechanically by breaking down the bigger structures into smaller metallic nanoparticles. In this case, nanoparticles purchased from the market are mixed with the base fluid to form a stable suspension. Therefore, the first step is the synthesis of nanoparticles and the second step is mixing the nanoparticles with base fluids. Next, the resultant mixture will undergo the sonication process. In this process, the mixture is stirred at very high ultrasonic frequencies for a considerable amount of time so that the suspensions become very stable and the nanoparticles are not sedimented. Other chemical methods exist in which nanoparticles are stabilized by modifying the pH value.

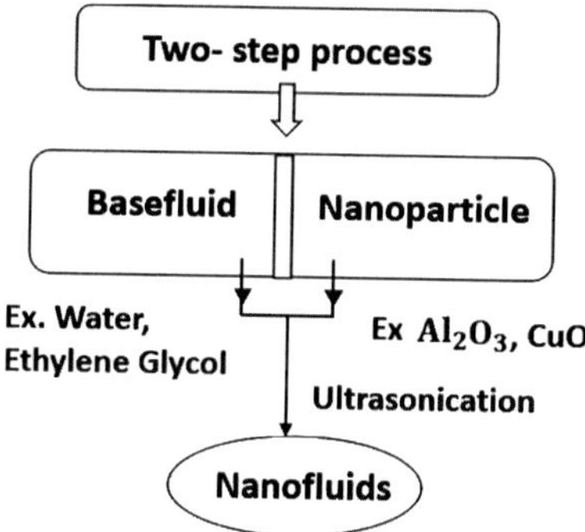

Fig. 8.2 Two-step process

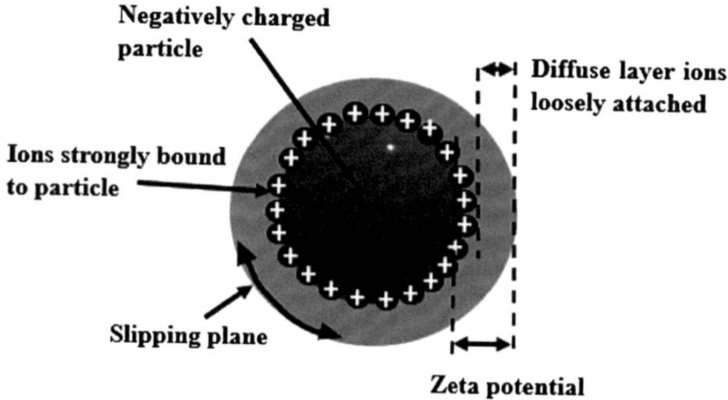

Fig. 8.3 Zeta potential representation of a nanoparticle

One of the most straightforward methods for checking the stability of nanoparticles is by measuring the zeta potential. Zeta potential is the excess potential that exists at the edge of the Stern layer, as shown in Fig. 8.3. The Zeta potential should be considerably high, that is, if the wall is positively charged, then the zeta potential can be greater than or equal to $+30$ mV, and if the wall is negatively charged, then the zeta potential can be less than or equal to -30 mV.

Therefore, zeta potential should be considerably large so that the particle near the wall cannot simply form an agglomerate with the particle on the wall. This large zeta electrostatic potential will repel these two particles because they have like charges. Therefore, the most important criterion to judge the stability of a nanofluid system is the zeta potential.

If the value of zeta potential is greater than or equal to $+30$ mV and less than or equal to -30 mV, then the suspension becomes quite stable. If it lies between -30 mV and 30 mV, then the nanoparticles are more likely to agglomerate.

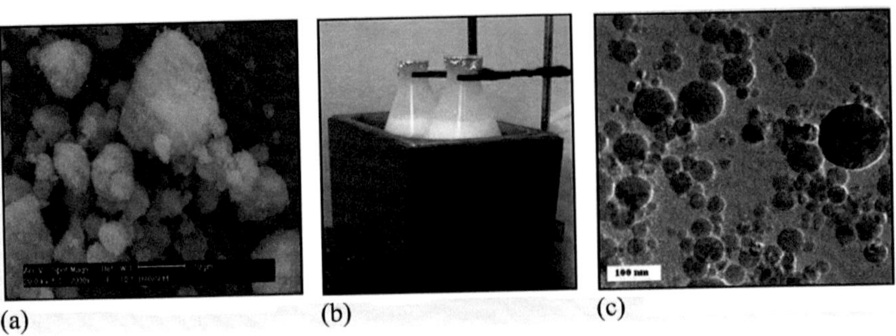

Fig. 8.4 **a** Before sonication; **b** an ultrasonic vibrator; **c** after sonication (Pawan K Singh, IITM, PhD thesis)

When the nanoparticles are dispersed in the base fluid, these particles form agglomerates, as shown in Fig. 8.4a. These are large agglomerates of nanoparticles. Typically, their values of zeta potential are quite low. Therefore, ultrasonic frequencies can be induced on these nanoparticles by subjecting them to the ultrasonic vibrator to increase the value of the zeta potential. This process is called sonication, which is shown in Fig. 8.4b.

In this process, all big clusters are broken and an excellent diffuse layer is formed, as a result of which the zeta potential increases considerably. This can be observed with the scanning electron microscope or the transmission electron microscope. These microscopes have a remarkably high resolution of the order of a few nanometers.

From these images, one can observe that the size of nanoparticles varies from a few nanometers to a few hundreds of nanometers. All the nanoparticles are not of the same size and shape. They have different size and shape distributions. However, the typical metallic oxides and nanoparticles are spherical (Fig. 8.5).

8.1.2 Effective Properties of Nanofluids

The effective properties of nanofluids can be calculated by considering the properties of nanoparticles and base fluids together. The effective properties such as specific heat, density, and volumetric thermal expansion coefficient are calculated using the following equations:

$$(\rho c_p)_{nf} = (1 - \phi)(\rho c_p)_f + \phi(\rho c_p)_s, \tag{8.1}$$

$$\rho_{nf} = (1 - \phi)\rho_f + \phi\rho_s, \tag{8.2}$$

$$\beta_{nf} = (1 - \phi)\beta_f + \phi\beta_s. \tag{8.3}$$

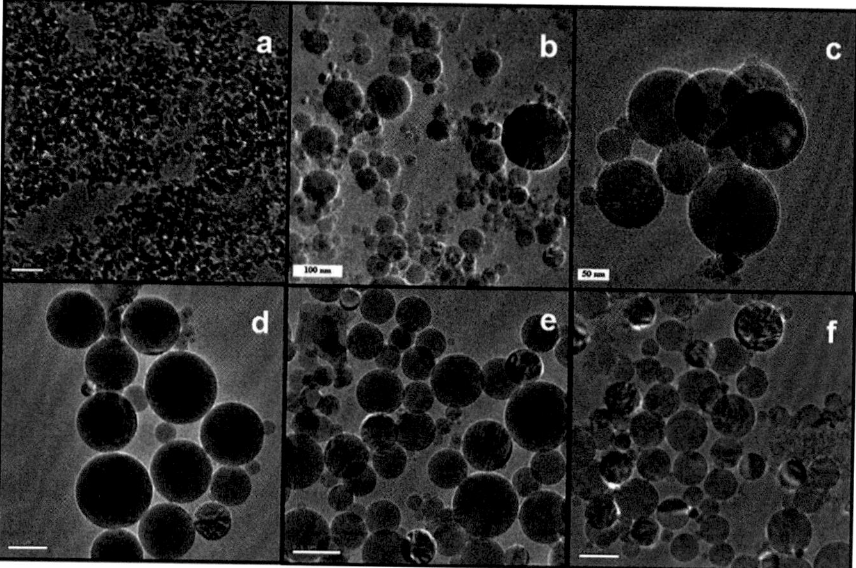

Fig. 8.5 Transmission electron microscope (TEM) images showing different sizes of nanoparticles

These effective properties are simply volume averaged. Other properties of nanofluids such as viscosity and thermal conductivity are measured experimentally. They cannot be volume averaged.

8.2 Measurement of Thermal Conductivity and Viscosity

Two rigorous techniques are available to determine the thermal conductivity and viscosity: the transient hot-wire method and the temperature oscillation method. The transient hot-wire method is one of the most popular methods for measuring the thermal conductivity of any fluid.

8.2.1 Transient Hot-Wire Method

The detailed method for measuring thermal conductivity is explained as follows. A long cylindrical container is filled with the nanofluid whose thermal conductivity has to be determined. At the center of the container, a very thin, long platinum wire is suspended, as shown in Fig. 8.6. The diameter of the wire is small and the length of the wire is relatively large, through which the current will be passed. One end of the platinum wire is connected to one of the resistances of a Wheatstone bridge.

212

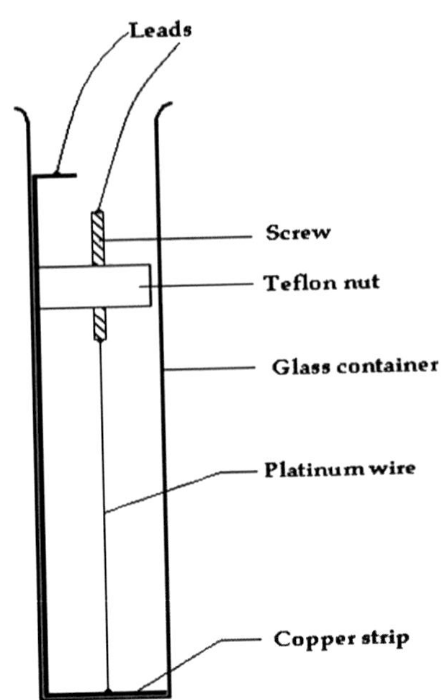

Fig. 8.6 Transient hot-wire measurement apparatus (Hrishikesh E. Patel, Thesis, Heat transfer in nanofluids)

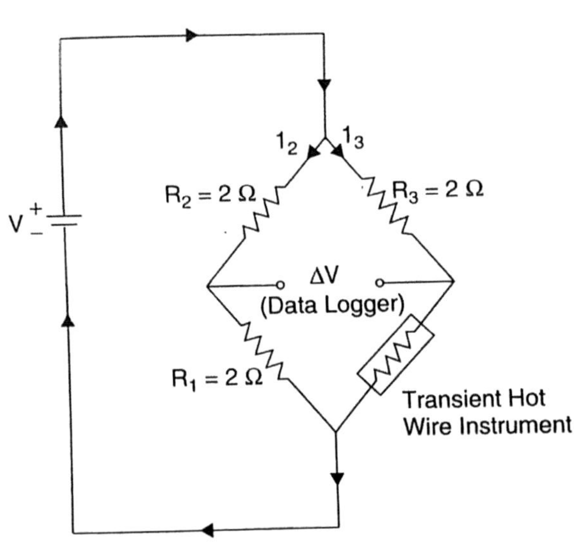

Fig. 8.7 Electric circuit for the transient hot-wire method (Hrishikesh E. Patel, Thesis-IITM, Heat transfer in nanofluids)

The Wheatstone bridge by default is a balanced bridge with four arms and four resistances, as shown in Fig. 8.7. There is no net voltage under balanced conditions. One of the arms gets unbalanced due to the change in resistance. On measuring the voltage, the unbalanced resistance can be calculated. This is the principle used in the transient hot-wire method. This comes under the transient heat conduction problem.

A platinum wire is usually chosen because of its higher temperature coefficient of resistance. While measuring thermal conductivity, a current is passed through the platinum wire. The dependence of electrical resistance on temperature is very high; it is quite sensitive to minute changes in temperature. This problem can be analyzed by writing the heat conduction equation in the cylindrical coordinate system. Heat diffuses radially from the platinum wire, which acts as a line heat source. Therefore, this system can be treated as a radially semi-infinite cylinder with a point line heat source. We can write the transient heat conduction equation assuming one-dimensional heat conduction in the radial direction of the surrounding nanofluid as follows:

$$\frac{\partial T}{\partial t} = \frac{\alpha}{r} \frac{\partial}{\partial r} \left(r \frac{\partial T}{\partial r} \right) + \frac{Q'''}{\rho C_p}. \tag{8.4}$$

The boundary conditions for this problem are as follows:

At any point on the wire, i.e., at $r = 0$, the value of temperature should be finite ($T = $ finite).

At $r = R$, the temperature at the other end is equal to the initial temperature ($T = T_{\text{initial}}$).

The expression for thermal conductivity (k) as a function of time and temperature can be obtained by applying the boundary conditions:

$$k = \frac{q}{4\pi (T_2 - T_1)} \ln \frac{t_2}{t_1}. \tag{8.5}$$

On measuring the temperature at two different time intervals, at time t_1, $T = T_1$ and after a certain time t_2, $T = T_2$. Based on these two values, the amount of heat supplied to the hot wire can be found out. From this, the thermal conductivity can be calculated.

Therefore, in this case, the platinum wire acts as both a heater and a sensor because from the change in the resistance in the Wheatstone bridge the actual value of resistance can be calculated. The Wheatstone bridge balances at a temperature where there is no liquid in the container and the hot wire is outside the container.

After suspending the wire in the fluid and passing the current, the wire starts cooling down. The cooling rate will be higher if the thermal conductivity of the fluid is high. This means the increase in temperature with respect to time will be lower. On plotting T versus $\ln t$, as shown in Fig. 8.8, the slope of this line will be smaller for larger thermal conductivities. Therefore, there is a change in the electrical resistance with respect to temperature.

Resistance is a function of time because the wire is cooling down while heat is being supplied. At the same time, heat is being dissipated to the fluid because of its thermal conductivity.

Therefore, the increase in temperature can be determined based on the thermal conductivity of the fluid. The temperature increases quite rapidly for a fluid with low thermal conductivity but increases gradually for a fluid with larger thermal

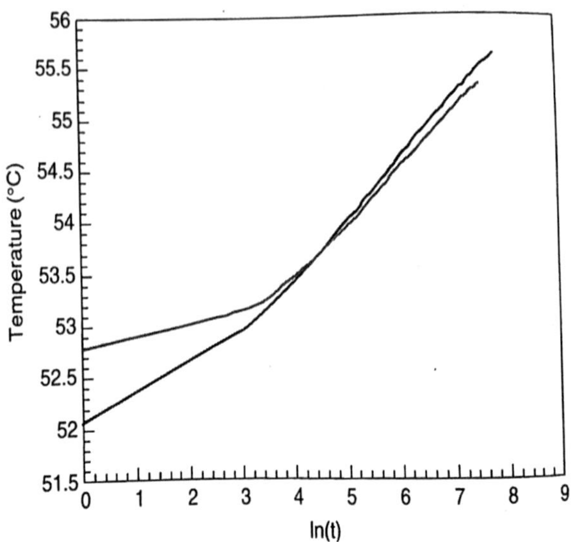

Fig. 8.8 Variation of temperature with time (Hrishikesh E. Patel, Thesis-IITM, Heat transfer in nanofluids)

conductivity. Therefore, on plotting T versus t at two different time intervals t_1 and t_2, the corresponding values of temperatures T_1 and T_2 are substituted into the expression for thermal conductivity. This process for measuring thermal conductivity is very fast. The values of temperature can be taken in 2–4 s. Therefore, natural convection effects can be ignored. This is one of the most common methods for measuring the thermal conductivity of any fluid using the transient process.

8.2.2 Temperature Oscillation Method

The other experimental technique available for measuring the thermal conductivity is the temperature oscillation method. The equipment used in this method is shown in Fig. 8.9. In this technique, the heater temperature changes as a function of the frequency of the sinusoidal wave. Therefore, an increase in temperature is correlated to the corresponding change in temperature. The effect of temperature on the fluid at the center of the container can be observed. The temperature of the fluid has to be measured at some arbitrary location inside the container and the temperature phase shift between the heater and the fluid at the center of the container has to be observed and correlated with thermal conductivity. This is called the temperature oscillation method.

The difference between the transient hot-wire method and the temperature oscillation method lies in the fact that the transient hot-wire method is a straightforward way of measuring thermal conductivity. The temperature oscillation method measures thermal diffusivity (α), and from its value, the heat capacity, density, and thermal

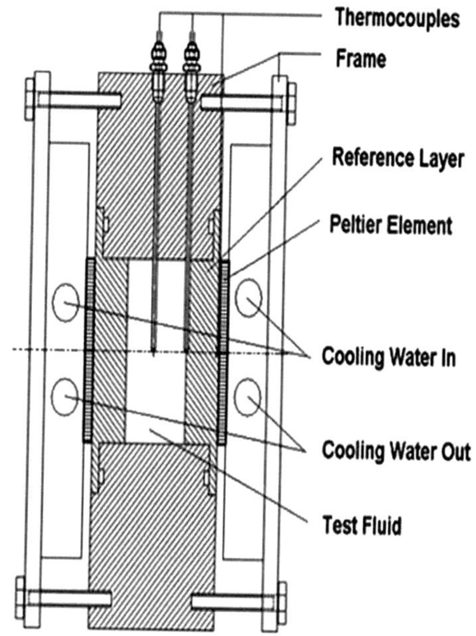

Fig. 8.9 Construction of the temperature-oscillation test cell (Hrishikesh E. Patel, Thesis-IITM, Heat transfer in nanofluids)

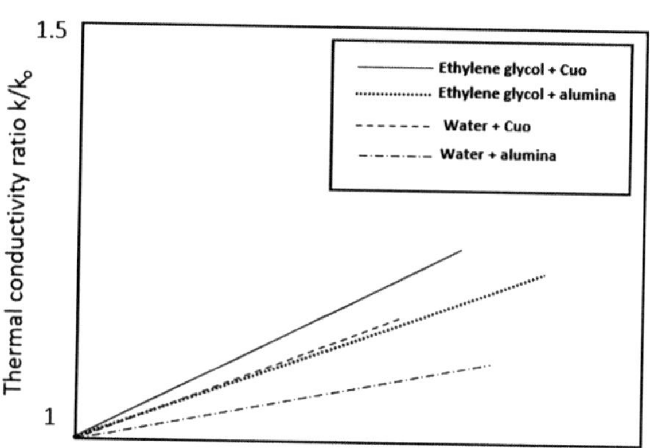

Fig. 8.10 Variation of thermal conductivity with volume fraction

conductivity are calculated. Therefore, this is a somewhat indirect way of measuring thermal conductivity. The measured value of thermal conductivity can be plotted as a function of volume fraction (ϕ). The volume fraction is the ratio of the volume of nanoparticles to the total volume. Typically, the nanoparticle volume fraction is low: 0.01, 0.02, and 0.03. The volume fraction values can be expressed in terms of %. Usually, ϕ should not exceed 5% because stable suspensions cannot be formed beyond 5%. Also, the pressure drop increases, which is not desirable.

Figure 8.10 shows the plot of the thermal conductivity ratio k/k_0 versus ϕ (where k/k_0 is the ratio of the thermal conductivity of the nanofluid to that of the base fluid)

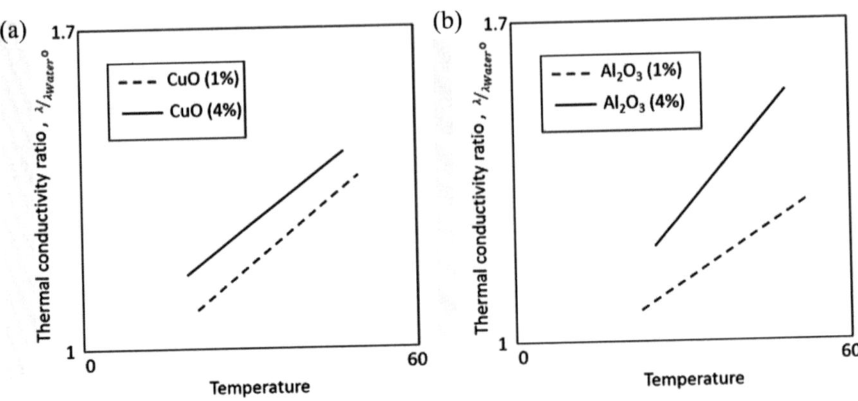

Fig. 8.11 Variation of thermal conductivity with temperature (°C) for **a** the CuO–water nanofluid and **b** the Al$_2$O$_3$–water nanofluid

for four different nanofluids: Al$_2$O$_3$–water, CuO–water, Al$_2$O$_3$–ethylene glycol, and CuO–ethylene glycol.

From the plot, the CuO–ethylene glycol nanofluid gives the highest enhancement in thermal conductivity. At 4%, the enhancement in thermal conductivity is 20%, whereas the alumina-water nanofluid gives not more than 10% enhancement. The enhancement entirely depends on the thermal conductivity of the nanoparticles and the base fluid. The thermal conductivity of ethylene glycol is lower than that of water. Therefore, the thermal conductivity of the Al$_2$O$_3$–ethylene glycol nanofluid is lower than that of the Al$_2$O$_3$–water nanofluid. Figure 8.11 shows the variation of thermal conductivity with temperature for Al$_2$O$_3$–water and CuO–water nanofluids. For a given volume fraction, on increasing the temperature of the nanofluid from 20° to 50°, a considerable enhancement in thermal conductivity from 20% to 40% can be observed for the CuO–water nanofluid and from 10% to 25% for the Al$_2$O$_3$–water nanofluid.

Thus, nanofluids have a very strong temperature dependence compared to base fluids. The thermal conductivity of liquids decrease with temperature. However, the thermal conductivity increases with temperature for nanofluids. This is where the nanoscale effects become relevant. One such effect is the Brownian motion. According to the Einstein model, when tiny particles are suspended as colloids, they will exhibit random motion in the base fluid, and the velocity of this motion is a function of temperature and is inversely proportional to the particle diameter and liquid viscosity. Therefore, as the diameter decreases, the temperature increases, which in turn increases the Brownian motion.

With the increase of Brownian motion, heat transport increases. Although the liquid appears static from a macroscale point of view, from a nanoscale point of view, nothing is static, and the nanoparticles are continuously moving in the base fluid. Therefore, the kinetic energy increases, which, in turn, contributes to the enhancement in thermal conductivity.

8.2.3 *Models Explaining Thermal Conductivity Enhancement in Nanofluids*

Different models explain the increase of thermal conductivity as a function of volume fraction. One of the most fundamental models is the Maxwell (classical model). This model gives the ratio of the effective thermal conductivity of the nanofluid to the thermal conductivity of the base fluid, i.e.,

$$\frac{k_{\text{eff}}}{k_f} = \frac{k_s + 2k_f - 2\phi(k_f - k_s)}{k_s + 2k_f + \phi(k_f - k_s)}, \tag{8.6}$$

where k_s is the thermal conductivity of nanoparticles, k_{bf} is the thermal conductivity of the base fluid, and ϕ is the nanoparticle volume fraction. From this model, it is clear that if

$$\phi = 0, \; k_{\text{eff}} = k_f$$

and if

$$\phi = 1, \; k_{\text{eff}} = k_s.$$

There are also some models to account for the shape of nanoparticles because the Maxwell model is only for spherical particles. Therefore, for nonspherical nanoparticles, the Hamilton–Crosser model (modified version of the Maxwell model) is widely used.

According to this model,

$$\frac{k_{\text{eff}}}{k_f} = \frac{k_s + (n - 1)k_f - (n - 1)\phi(k_f - k_s)}{k_s + (n - 1)k_f + \phi(k_f - k_s)}, \tag{8.7}$$

where n is the nanoparticle shape factor, $n = \frac{3}{\psi}$, where ψ is the sphericity of particles. For example, ψ is 1 for spherical particles. Therefore, the expression for the Maxwell model can be obtained by substituting $n = 3$ in the Hamilton–Crosser model.

The Davis model is another model used for calculating the effective thermal conductivity of nanofluids. This model includes some higher order terms, which represent the interaction between the randomly dispersed spherical nanoparticles. By neglecting the second- and third-order terms, this model becomes the Maxwell model:

$$\frac{k_{nf}}{k_{bf}} = 1 + \frac{3\left(k_s/k_{bf}\right) - 1}{\left(k_s/k_{bf} + 2\right) - \left(k_s/k_{bf} - 1\right)\phi}\left(\phi + f\left(k_s/k_{bf}\right)\phi^2 + O(\phi^3)\right). \tag{8.8}$$

All these constitute the fundamental models widely used to predict the effective thermal conductivity of nanofluids as a function of the volume fraction, but an important behavior is the variation of temperature. This aspect has not been included in these models. The base fluid shows a reduction in thermal conductivity with temperature.

Nevertheless, the nanofluid shows a continuous increase. This can be predicted or modeled only by accounting for the Brownian motion. The Brownian motion is the most important mechanism describing the phenomenon involved in the enhancement of thermal conductivity.

One thermal conductivity model was developed recently by Dhar et al. [32]. This model is the most comprehensive one, and accounts for two components: random (Brownian) motion and percolation. This model was developed particularly for Graphene nanofluids. The thermal conductivity of nanofluids can be estimated according to this model:

$$k_{gnf} = k_{sd}^{\alpha} . k_{perc}^{1-\alpha}. \tag{8.9}$$

The first component is called sheet dynamics (k_{sd}). The sheet dynamics part is the summation of three components: (1) the thermal conductivity of the base fluid (k_{medium}), (2) the thermal conductivity caused by the Brownian motion of nanoparticles ($k_{dynamics}$), and (3) the thermal conductivity of the fluid solely because of the nanosheets (k_{EMT}). Therefore,

$$k_{sd} = k_{medium} + k_{EMT} + k_{dynamics}, \tag{8.10}$$

where k_{EMT} can be expressed as

$$k_{EMT} = \frac{k_p \phi d_m}{(1 - \phi) L_g}, \tag{8.11}$$

where k_p is the thermal conductivity of graphene, d_m is the molecular diameter of the base fluid, ϕ is the nanoparticle volume fraction, and L_g is the effective face size of nanosheets. L_g is of the order of a few hundreds of nanometers, whereas d_m is of the order of a few angstroms. Therefore, if ϕ tends to 1, the value of k_{EMP} approaches the value of k_p. $k_{dynamics}$ is the dynamic thermal conductivity due to the Brownian motion and is given as

$$k_{dynamics} = \bar{U}_B \lambda C_V \phi \theta, \tag{8.12}$$

where Λ is the mean free path, which is assumed to be 10^{-6} m, and C_v is the volumetric specific heat, which is given as

$$C_V = \frac{36\pi^4 k_B}{15} \left(\frac{N}{V} \right) \left(\frac{T}{\theta_D} \right)^3, \tag{8.13}$$

where $\frac{N}{V}$ is the number of atoms per unit volume and θ_D is the Debye temperature of graphene for planar modes of phonon transport. Brownian velocity obtained from the kinetic theory of gases is given by

$$\bar{U}_B = 3U_B, \tag{8.14}$$

where the Brownian velocity is

$$U_B = \frac{2k_B T}{\pi \mu_{bf} L_G^2}.$$ (8.15)

It is thus a function of the Brownian velocity U_B, volumetric heat capacity, nanoparticle volume fraction, Debye temperature, and mean free path.

It is also called the Einstein scope expression. This shows that the Brownian velocity is directly proportional to temperature and inversely proportional to the particle size. Therefore, with this model, the variation of thermal conductivity with temperature can be predicted accurately. This model got wide acceptance and gave outstanding results for different nanofluids, including both particle-type and sheet-type nanofluids.

The other vital component that contributes to the calculation of thermal conductivity of nanofluids is the percolation component. This component is crucial for sheet-type nanofluids such as carbon nanotubes. Nanosheets form linkages with several other sheets, which are linked together in a chain, and several of these chains exist in parallel. These chains contribute to the conduction of heat from the left side to the right side because they form a conduction network, which promotes the transport of heat.

Consider a unit cube of nanofluid in which nanoparticles are dispersed, as shown in Fig. 8.12. The nanoparticles can be generic and can be of any shape and cross-section. In the generic case, they can be like sheets connected with each other. One cannot find the linkage structure in the case of pure particles. They are just dispersed in the base fluid. These nanosheets most likely form chains. The length of any microsheet is equal to the face size of these sheets.

To understand the percolation component, the number of parallel chains that can be formed has to be estimated. The total number of parallel graphene chains in the domain is given by

$$M = \frac{\phi L_{cell}^3}{\left(n_g - 1\right) d_g L_g^2 N},$$ (8.16)

where ϕ is the volume fraction, n_g is the average number of layers of graphene nanosheets, d_g is the intersheet distance for the graphene sample, and N is the number of nanosheets in a single percolation chain. If the number of nanosheets in a single chain is known, one can use this empirical correlation to find out the total number of chains that can be formed.

The resistance network analogy can be used to obtain the total thermal resistance in the nanofluid. For this, the number of resistances acting in the system should be known. The conduction resistance acts through the nanosheet and interfacial thermal resistance acts between two parallel nanosheets. Note that the nanosheets are not precisely touching each other. A small gap exists between successive nanosheets, which are separated by the nanofluid. There exists some contact resistance between the fluid and the solid on either end of the nanoparticle. Therefore, the conduction

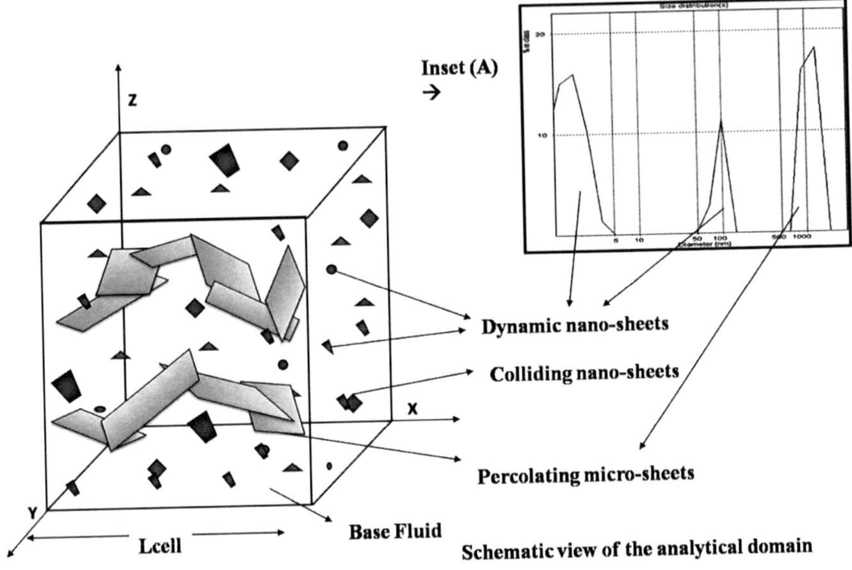

Fig. 8.12 Unit cube of nanofluid [32]

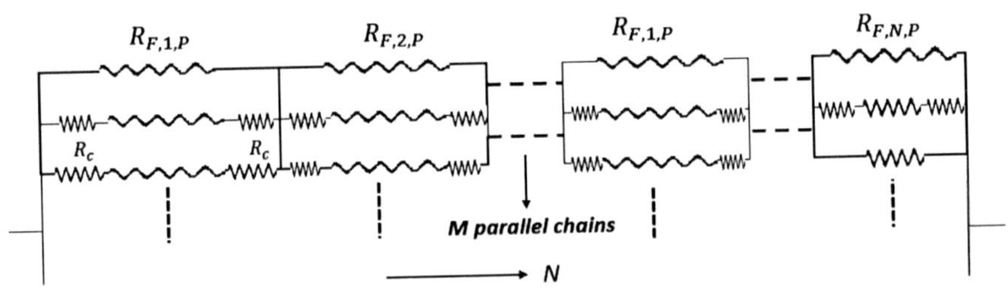

Fig. 8.13 The net resistance approach

resistance of the nanoparticle and twice the contact resistance on either end act on a single nanosheet. This is the series component of the resistance. There exist N such nanoparticles in a single chain. A parallel component of resistance is present, which is the heat conduction through the base fluid. There are M such chains in parallel.

The total thermal resistance acting in this system can be obtained by drawing the resistance network analogy, as shown in Fig. 8.13. Therefore, the net effective resistance is given by

$$R_{net} = \sum_{i=1}^{N} \left[\frac{\left(\frac{R_{ip}+2R_C}{M} \right) \times R_{Fi,p}}{\left(\left(\frac{R_{ip}+2R_C}{M} \right) + R_{Fi,p} \right)} \right], \tag{8.17}$$

where $R_{Fi,p}$ is the conduction resistance through each nanoparticle and R_c is the interfacial contact resistance between two successive nanosheets given by

$$R_c = \frac{1}{A_{contact}\,G} = \frac{1}{\left(n_g - 1\right) d_g L_g G},\tag{8.18}$$

where G is the thermal conductance per unit area. $G = 25\,MWm^{-2}K^{-1}$ for the graphene–water interface and $G = 12\,MWm^{-2}K$. $R_{i,p}$ is the thermal resistance of the graphene nanosheet. Therefore, from Fourier's law of heat conduction, the thermal conductivity based on percolation theory is given by

$$k_{\text{perc}} = L_{\text{cell}}/\left(R_{\text{net}} A_{\text{cell}}\right),\tag{8.19}$$

where L_{cell} is the length of the cell and A_{cell} is the cross-sectional area of the cell.

The total thermal conductivity of the nanofluid can be calculated from both percolation and sheet dynamics components using the power law.

$$k_{gnf} = k_{sd}^{\alpha}.k_{perc}^{(1-\alpha)}.\tag{8.20}$$

The factors α and $1 - \alpha$ represent the individual contributions of sheet dynamics and percolation components, respectively. α is the sheet distribution factor, which signifies the amount of weightage to be given for both sheet dynamics and percolation. If the nanofluid consists of only spherical particles ($\alpha = 1$), then the percolation component becomes 0. Therefore, only the Brownian motion is responsible for the increase in thermal conductivity. However, in the case of carbon nanotubes, flake structures and linkages will be formed. Therefore, the Brownian motion component is almost 0 because they cannot move as fast as the spherical particles. In this case, α becomes 0. Only the percolation component contributes to the total effective conductivity. In the case of the graphene nanofluid, both Brownian motion and conduction components contribute to the effective thermal conductivity. In this case, the value of α may be 0.7 or 0.8. Figure 8.14 shows the % enhancement in the effective thermal conductivity with α for different nanoparticle volume fractions.

The value of α for graphene is 0.34 and $\alpha = 1$ for pure spherical metallic nanoparticles such as alumina and copper. This factor (α) is not an adjustable parameter in the model but can be theoretically determined. It can be proclaimed that if the particle setting velocity under gravity of a particle is more than the Brownian velocity, it can form the percolation chain while the particle will be in Brownian motion if the Brownian velocity is higher. Since, the particle size distribution can be measured directly, following the above principle, the fraction in percolation (α) can be calculated. This is a comprehensive model which accounts for the reason why and how the thermal conductivity increases with temperature and volume fraction for not only spherical nanoparticles but also nanoparticles of different shapes. This is one of the best models available in the literature for different nanofluids. This model shows satisfactory agreement with the experimental results.

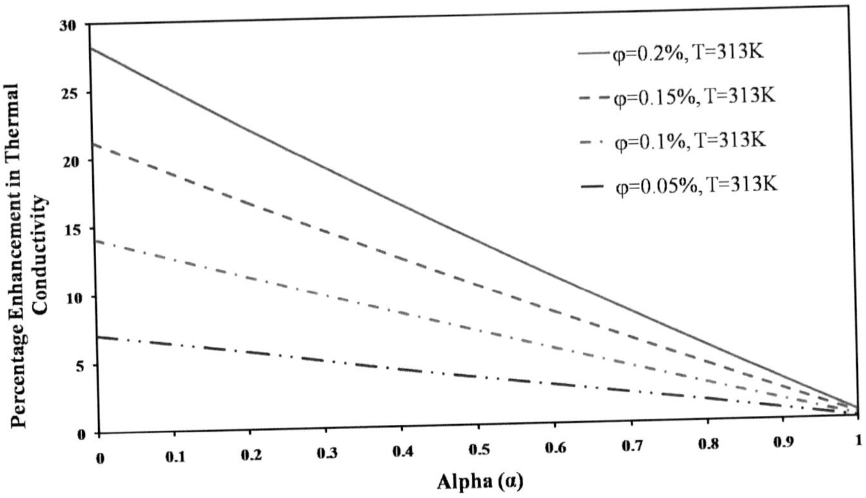

Fig. 8.14 Percentage enhancement in effective thermal conductivity with α for different volume fractions [32]

8.2.4 *Measurement of the Viscosity of Nanofluids*

The addition of nanoparticles to base fluids increases their viscosity. The effective viscosity of a nanofluid cannot be calculated using the mixture rule. Therefore, a viscometer can be used to estimate the dynamic viscosity of nanofluids. Using a rheometer, it can be identified whether a nanofluid is Newtonian or non-Newtonian. A viscometer and a rheometer are shown in Fig. 8.15. Therefore, a rheometer estimates not only the dynamic viscosity but also the type of nanofluid. For a given shear stress, the corresponding strain rate can be determined using the rheometer. From the stress versus strain curve, one can identify whether the plot is linear or nonlinear so that the given fluid can be classified as a Newtonian fluid or a non-Newtonian fluid. Therefore, the rheometer can be considered as a better tool to assess the fundamental Newtonian behavior of nanofluids. Most of the nanofluids such as alumina, copper, carbon nanotubes, and graphite behave like Newtonian fluids.

Figure 8.16 shows the variation of viscosity with nanoparticle volume fraction. From the figure, viscosity increases with nanoparticle volume fraction (ϕ). The effect of nanoparticle diameter on percentage increases in viscosity is shown in Fig. 8.17. From the figure, it is clear that viscosity increases with nanoparticle diameter. The influence of temperature on viscosity is shown in Fig. 8.18. From this figure, one can observe that viscosity decreases with an increase in temperature.

Fig. 8.15 a A viscometer; **b** a rheometer

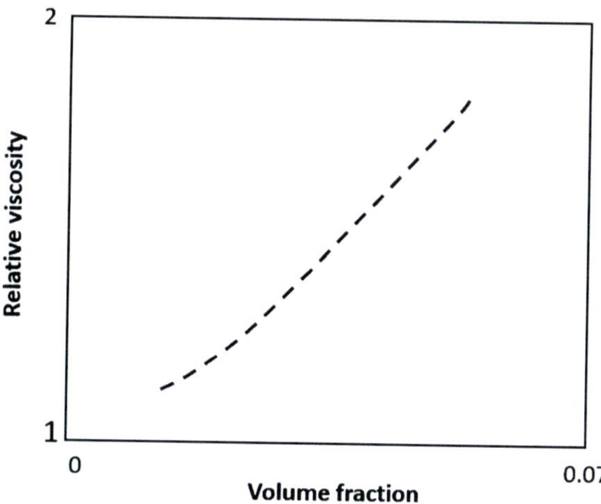

Fig. 8.16 Variation of viscosity with nanoparticle volume fraction

Several models can predict the viscosity of nanofluids concerning both nanoparticle volume fraction and temperature. They are given as follows:

The Einstein model [35]:

$$\frac{\mu_{nf}}{\mu_f} = 1 + 2.5\phi.$$

$$(8.21)$$

The Batchelor model (1972):

$$\frac{\mu_{nf}}{\mu_f} = 1 + 2.5\phi + 6.5\phi^2.$$

$$(8.22)$$

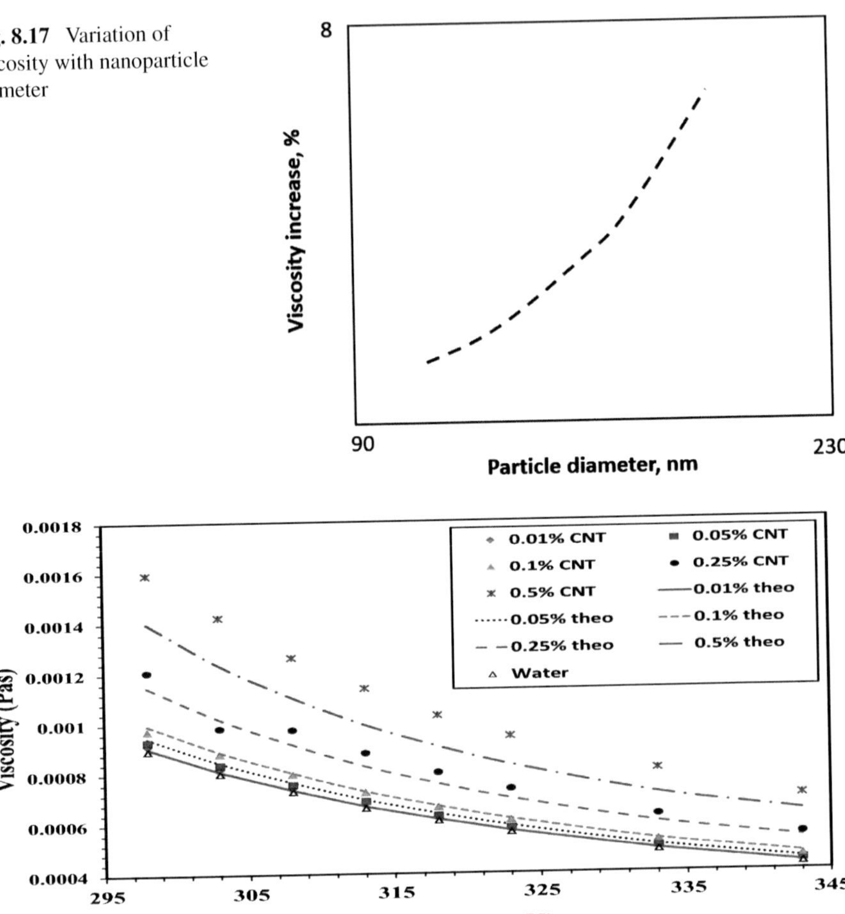

Fig. 8.17 Variation of viscosity with nanoparticle diameter

Fig. 8.18 Variation of viscosity with temperature [33]

The Chandrashekar model:

$$\frac{\mu_{nf}}{\mu_f} = 1 + b\left(\frac{\phi}{1 - \phi_m}\right)^n. \tag{8.23}$$

The Abu-Nada model:

$$\mu_{nf} = -0.155 - \frac{19.582}{T} + 0.794\phi + \frac{2094.47}{T^2} - 0.192\phi^2 - 8.11\frac{\phi}{T}$$
$$- \frac{27463.863}{T^3} + 0.127\phi^3 + 1.6044\frac{\phi^2}{T} + 2.1754\frac{\phi}{T^2}. \tag{8.24}$$

Dhar et al. [32] proposed a model using both percolation and sheet dynamics components for graphene nanosheets. These components affect not only the thermal conductivity but also the viscosity. According to this model, the viscosity of nanofluids can be expressed as

$$\mu_{GNS} = \mu_{perc} + \mu_{sd}. \tag{8.25}$$

The component μ_{perc} is the viscosity due to sheet percolation:

$$\mu_{perc} = \mu_{bf}\left(1 + L^*d\phi\alpha\right), \tag{8.26}$$

where μ_{bf} is the viscosity of the base fluid, the non-dimensional length $L^* = L_G/L_{crit}$.

L_G is the average sheet face size of graphene and L_{crit} is the critical sheet size, d is the percolation network dynamicity factor, ϕ is the nanoparticle volume fraction, and α is the sheet distribution.

The value of L^* for graphene is approximately equal to 1.25. The component μ_{sd} is the viscosity due to sheet dynamics, which is given by

$$\mu_{sd} = \mu_0\phi\left(1 - \alpha\right), \tag{8.27}$$

where μ_0 is the dynamic viscosity term, which is given by

$$\mu_0 = \rho_G\Lambda U_B\theta, \tag{8.28}$$

where ρ_G is the density of the graphene, Λ is the mean free path, U_B is the Brownian velocity, and θ is the collision cross-section.

The dynamic viscosity of carbon nanotubes (CNTs) is given by

$$\mu_{CNTS} = \mu_{bf}\left(1 + L^*d\phi\right). \tag{8.29}$$

The value of L^* for CNTs is approximately equal to 10.

Figure 8.19 shows the variation of viscosity with nanoparticle volume fraction for graphene–water and CNT–water nanofluids for various temperatures. From these figures, one can observe that the viscosity increases with the nanoparticle volume fraction. The Dhar et al. (2013) model can also very accurately predict the dependence of the effective viscosity of graphene–water nanofluids and CNT–water nanofluids on nanoparticle volume fraction and temperature.

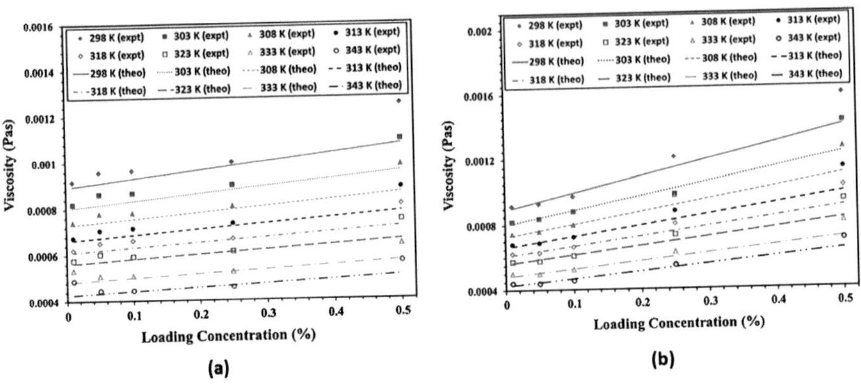

Fig. 8.19 Variation of viscosity with volume fraction for **a** graphene–water and **b** CNT–water nanofluids [33]

8.3 Heat Transfer in Nanofluids

So far, the thermophysical properties of nanofluids and their measurement techniques have been discussed in detail for a static fluid. Several experiments have been conducted on the single-phase forced convection, natural convection, and phase change phenomenon by replacing base fluids with nanofluids. The forced convection experiment is explained as follows.

In the given setup, the tube is heated by using a DC power supply and the base fluid is passed through the tube. The average Nusselt number for the base fluid is calculated by measuring the heat flux and the temperature distribution along the wall of the tube. Afterward, the base fluid is replaced with the nanofluid and Nu_{avg} is calculated. The ratio of the Nusselt number of the nanofluid to the Nusselt number of the base fluid is plotted, and the enhancement in Nu_{avg} is observed by using the nanofluid. The results clearly show that the enhancement is caused by not only the thermal conductivity of the nanoparticles but also the convection component.

The average heat transfer coefficient can be expressed as

$$h = \frac{Q''}{(T_w - T_m)} = \frac{-k_{nf} \left. \frac{\partial T}{\partial y} \right|_{y=0}}{(T_w - T_m)}, \tag{8.30}$$

where T_w is the wall temperature and T_m is the bulk mean temperature. Therefore, the average Nusselt number can be expressed as

$$Nu = \frac{hx}{k_{nf}}. \tag{8.31}$$

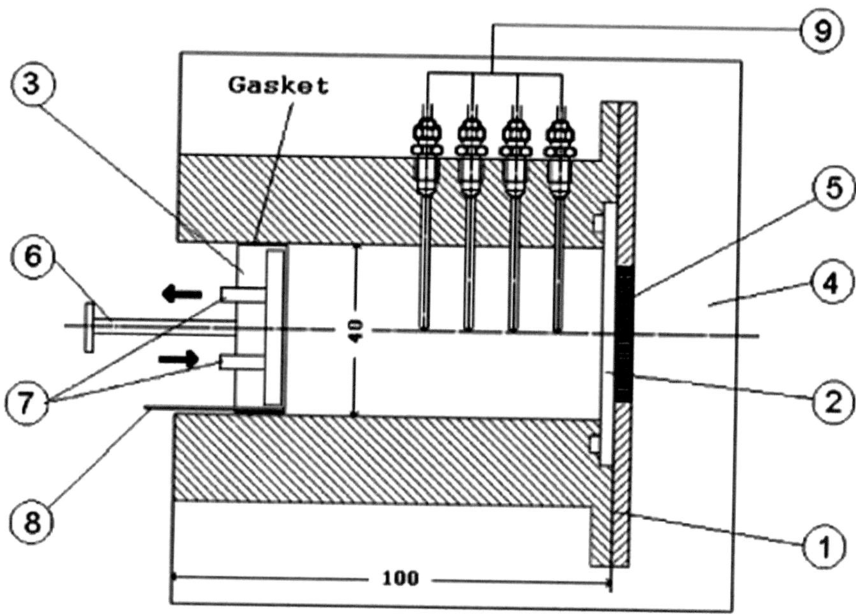

1. **Cylindrical block**
2. **End cover as heating surface.**
3. **End cover as cooling surface**
4. **Cap**
5. **Resistance heating elements**

6. **The piston shaft**
7. **Cooling water inlet and outlet**
8. **Narrow tube**
9. **Thermocouples**

Fig. 8.20 Natural convection experimental setup [80]

On substituting h in the expression for the Nusselt number, the thermal conductivity k_{nf} cancels and the resultant expression becomes

$$Nu = \frac{\left.\frac{\partial T}{\partial y}\right|_{y=0} x}{(T_w - T_m)}. \tag{8.32}$$

Therefore, the increase in the Nusselt number on replacing the base fluid with the nanofluid is mainly because of the increase in the temperature gradient, and not because of the increase in thermal conductivity. The value of heat transfer coefficient (h) can be higher for nanofluids but the effect of the thermal conductivity increase can be compensated by using k_{nf} in the expression for the Nusselt number. The Nusselt number for nanofluids is related only to the temperature gradient. The increase in temperature gradient results in better heat transfer characteristics of nanofluids. This has been demonstrated clearly by Anoop et. al where the increase in heat transfer coefficient has been shown in Fig. 8.20a and that of Nusselt number is shown in

Fig. 8.20b. The correlations suggested by Anoop et al. are as follows. (should include the figures and correlations).

Some standard correlations for average Nusselt numbers are available for nanofluids in the laminar regime. Xuan and Li [97] proposed a correlation for the average Nusselt number for the entire tube as a function of Reynolds number, Prandtl number, Peclet number, and nanoparticle volume fraction. This can be expressed as

$$Nu = 0.4328(1 + 11.285\varepsilon_p^{0.754} Pe_d^{0.218})Re^{0.3333} Pe^{0.4}. \tag{8.33}$$

Maiga et al. [74] proposed two different correlations for a Nusselt number as a function of Reynolds number and Prandtl number for both constant heat flux and constant wall temperature:

$$\overline{Nu} = 0.086 Re^{0.55} Pr^{0.5}, \tag{8.34}$$

$$\overline{Nu} = 0.28 Re^{0.35} Pr^{0.36}. \tag{8.35}$$

These correlations are valid for $Re \leq 1000$ and $6 \leq Pr \leq 753$.

The increase in average Nusselt number is not only because of the effective thermal conductivity of nanofluids but also due to the convective heat transfer enhancement. The thermal conductivity enhancement contributes to conduction heat transfer, but the principal focus is only on the temperature gradient, which is a measure of the convective heat transfer enhancement. Apparently, most of the experimental studies agree that there is an increase in the average Nusselt number with the addition of nanoparticles, but only the percentage increase is not very consistent.

Similarly, in the turbulent regime, there is a possible enhancement with the addition of nanoparticles. Experimental correlations for the average Nusselt number in turbulent flow are available in the literature for flow in a circular tube (Fig. 8.21).

Pak and Cho [78] proposed a correlation for the average Nusselt number, called the modified Dittus–Boelter equation, i.e.,

$$Nu = 0.021 Re^{0.8} Pr^{0.5}. \tag{8.36}$$

Xuan and Li [97] proposed a correlation for the average Nusselt number, which is given by

$$Nu = 0.0059 \left(1 + 7.6286\varepsilon_p^{0.6886} Pe_d^{0.001}\right) Re^{0.9238} Pr^{0.4}, \tag{8.37}$$

where Pe_d is the Peclet number of particles and ε_p is the volume fraction of the particles.

Putra et al. [80] conducted an experiment on the natural convection of a nanofluid in a cavity and found that there was deterioration of the Nusselt number in the natural convection regime. The experimental setup is shown in Fig. 8.20. With the addition of high-density nanoparticles to the base fluid, the effective density of the nanofluid increases. Therefore, the buoyancy force decreases. Lighter fluids can experience

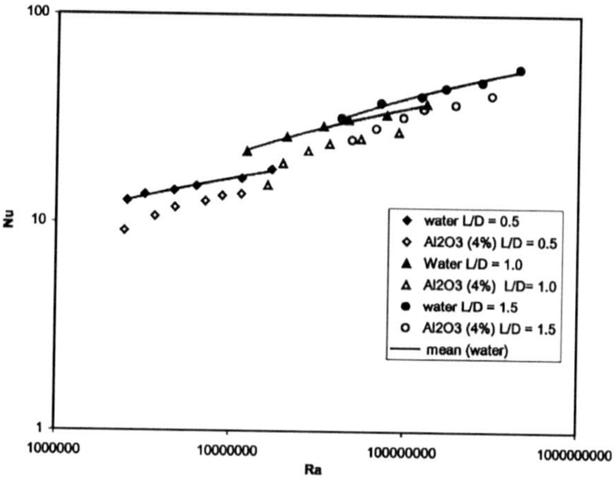

Fig. 8.21 Variation of the Nusselt number with the Rayleigh number for different aspect ratios [80]

the natural convection phenomenon better compared to heavier fluids. Hence, there is deterioration of heat transfer in the natural convection regime. Figure 8.21 shows the variation of the Nusselt number with the Rayleigh number for different aspect ratios.

However, in the case of carbon nanotubes (CNTs) and flake-type nanoparticles, heat transfer enhancement can be observed in the natural convection regime. On plotting the normalized Nusselt number versus the base fluid, the alumina–water nanofluid shows that for all the Rayleigh numbers, the normalized Nusselt number is always less than that of the base fluid or approximately equal to that of the base fluid, as shown in Fig. 8.22.

The MWCNT–water nanofluid and the graphene–water nanofluid show an enhancement of 30–35% at $Ra = 10^6$. In general, the addition of nanoparticles contributes to the overall increase in density, which results in the deterioration of buoyancy force. Therefore, heat transfer should decrease. However, an enhancement of heat transfer is observed on the addition of CNTs and graphene. This phenomenon cannot be explained simplistically. Researchers performed a scaling analysis to explain this phenomenon and found that apart from Brownian force several other forces such as drag force also play a vital role in heat transfer enhancement. The drag force makes a significant contribution especially in the case of flake-type nanoparticles, carbon nanotubes, and graphene compared to spherical nanoparticles.

Therefore, natural convection is augmented using drag force. This is the only way one can explain why specific nanoparticles such as carbon nanotubes and graphene can result in a heat transfer enhancement. However, according to the general theory, heavier nanoparticles present in the base fluid should contribute to deterioration of heat transfer.

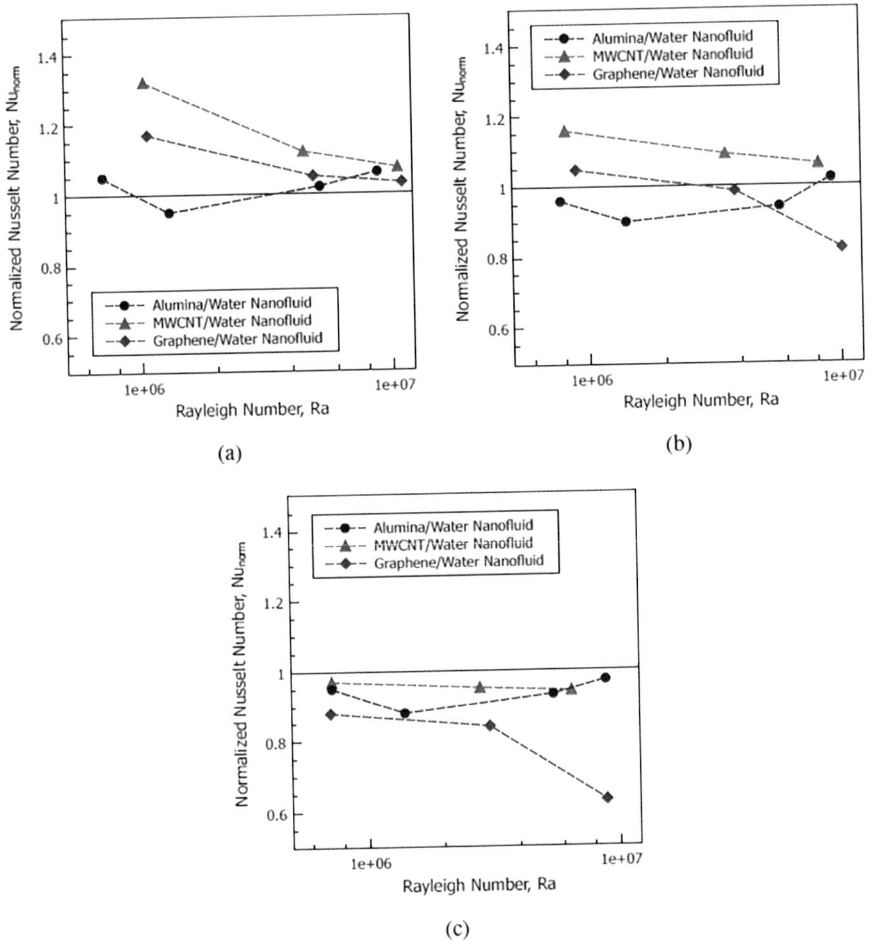

Fig. 8.22 Normalized Nusselt number versus Rayleigh number for **a** $\phi = 0.1$; **b** $\phi = 0.3$; **c** $\phi = 0.5$; **d** for all the volume fractions

8.4 Boiling of Nanofluids

The boiling characteristics of nanofluids are studied by using pool boiling experiments. The experimental setup is shown in Fig. 8.23.

The heat transfer coefficient is found to decrease with an increase in the volume fraction of nanoparticles. Therefore, the Nusselt number value deteriorates in pool boiling.

Das et al. [31] developed heat transfer correlations for smooth and rough heating surfaces, as shown in Tables 8.1 and 8.2, where Re_b is the boiling Reynolds number, which can be defined as the ratio of the bubble inertial force to the liquid viscous force. It differs from the flow Reynolds number. It is based on the velocity of nucle-

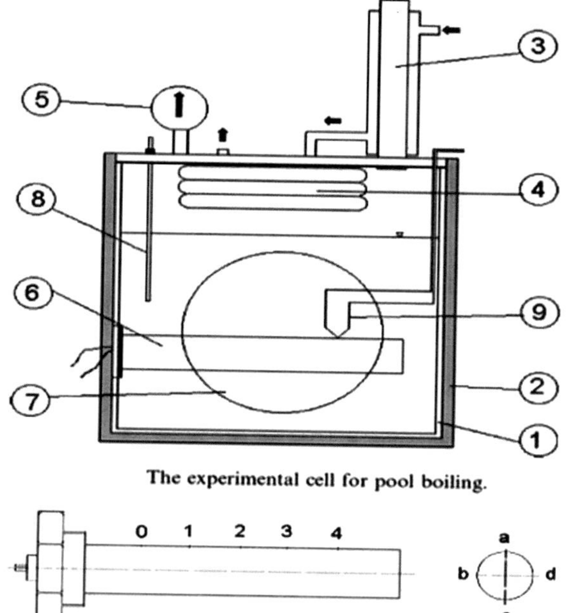

The experimental cell for pool boiling.

The cartridge heater with thermocouple locations.

Fig. 8.23 Pool boiling experimental setup [30]

Table 8.1 Heat transfer correlations for a smooth heating surface [31]

Pure water	$Nu = 97.9\mathrm{Re}_b^{0.638}\mathrm{Pr}^{0.4}$
1%	$Nu = 78.84\mathrm{Re}_b^{0.687}\mathrm{Pr}^{0.4}$
2%	$Nu = 72.39\mathrm{Re}_b^{0.69}\mathrm{Pr}^{0.4}$
4%	$Nu = 67.56\mathrm{Re}_b^{0.619}\mathrm{Pr}^{0.4}$

Table 8.2 Heat transfer correlations for a rough heating surface [31]

Pure water	$Nu = 137\mathrm{Re}_b^{0.526}\mathrm{Pr}^{0.4}$
1%	$Nu = 99.48\mathrm{Re}_b^{0.503}\mathrm{Pr}^{0.4}$
2%	$Nu = 94.63\mathrm{Re}_b^{0.495}\mathrm{Pr}^{0.4}$
4%	$Nu = 89.12\mathrm{Re}_b^{0.49}\mathrm{Pr}^{0.4}$

ating bubbles and the bubble diameter. From these correlations, it is found that a rough heating surface will have a higher value of Nusselt number compared to a smooth one. The multiplying coefficients are larger in the case of a rough heating surface because of the overall higher heat transfer coefficient as roughness promotes additional nucleation sites, which thereby improves the nucleate boiling capability.

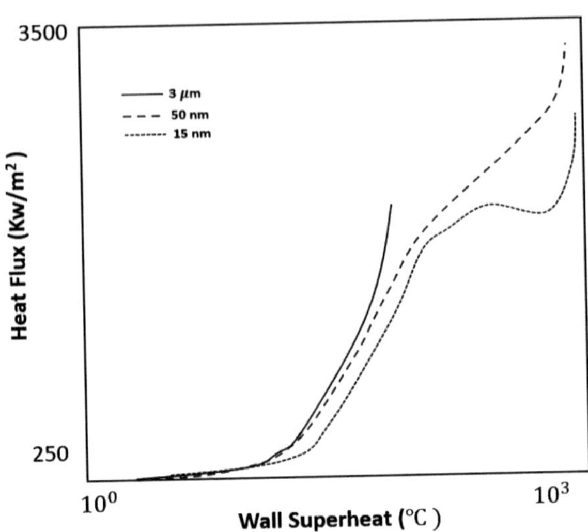

Fig. 8.24 Typical Boiling curves of an NiCr wire ($D = 0.4$ mm) in silica–water solutions

A critical aspect of nanofluids from a cooling point of view is that the critical heat flux is enhanced by the addition of nanoparticles. This is very important for nuclear safety. On increasing the value of the critical heat flux, the chances of transition from nuclear boiling to film boiling will decrease. This is a favorable condition for preventing dry-out, which leads to no further increase in the vapor pressure and reduces the explosion chances.

Vassallo et al. [95] carried out pool boiling experiments for various volume fractions and different diameters of silica particles. They observed an enhancement in the critical heat flux (CHF) of about 200% for silica–water nanofluids compared to water for 15- and 50-nm nanoparticles shown in Fig. 8.24.

When silica particles of about 3 µm diameter were suspended in water, the enhancement in CHF was found to be only 100%. For this diameter of silica, the nanoparticles wire failed before attaining film boiling.

Not many studies have been reported on the flow boiling of nanofluids. However, in flow boiling, the critical heat flux can be augmented with the addition of nanoparticles by varying the Reynolds number.

Exercise Problems

1. Properties of the basefluid at room temperature are given below: Properties of the nanoparticles are given below (Table 8.2):

   Consider nanofluids: Alumina/Water, MWCNT/Water and Graphene/Water, Alumina/Ethylene Glycol, MWCNT/ Ethylene Glycol and Graphene/ Ethylene Glycol.

  a. Calculate the effective thermophysical properties such as ρ_{eff}, $c_{p,eff}$, μ_{eff}, and k_{eff} for all above nanofluids for volume fractions of 0.1%, 0.5%, 1%, 2%, 3%, and 4% at room temperature.

  b. Plot the thermophysical properties as the function of above given volume fractions.

  c. Compare k_{eff} values for above given volume fractions by (a) Maxwell Model, (b) Hamilton–Crosser Model. (Use $\psi = 1$ for alumina, $\psi = 0.5$ for MWCNT, $\psi = 0.6$ for Graphene nanoparticles.)

  d. Compare eff values for above given volume fractions by (i) Einstein Model, (ii) Batchelor Model, and (iii) Abu-Nada Model.

2. Consider Alumina/Water nanofluid following through the circular tube of diameter 5 mm. The temperature at the inlet and outlet is $25\,°C$ and $35\,°C$ respectively. The mass flow rate is $4 \times 10^{-2}\,kg/m^3$. The diameter of Alumina nanoparticle is 80 nm.

Properties	Water	Ethylene Glycol
Density (ρ)	995.7 (kg/m³)	1110 (kg/m³)
Specific heat (c_p)	4.178 (kJ/(kgK))	3.12 (kJ/(kgK))
Thermal conductivity (k)	7.98e(–04) (Pa.s)	1.02e(–02) (Pa.s)
Dynamic viscosity (μ)	0.615 (W/Km)	0.258 (W/Km)

Properties	Alumina	MWCNT	Graphene
Density (ρ)	3950 (kg/m³)	2000 (kg/m³)	1100 (kg/m³)
Specific heat (c_p)	765 (J/(kgK))	710 (J/(kgK))	710 (J/(kgK))
Thermal conductivity (k)	40 (W/Km)	2800 (W/Km)	3000 (W/Km)

  a. Calculate the Nusselt number for volume fraction of 0.1%, 0.5%, 1%, 2%, 3%, and 4% using Xuan and Li correlation [97]. Use Maxwell model for finding effective thermal conductivity and Einstein model to find effective viscosity of nanofluid. All other properties are given in the Tables 8.1 and 8.2. Peclet number based on the particle diameter is given by

$$Pe_d = \frac{u_m \, d_f}{\alpha_{nf}} \tag{8.38}$$

$$\alpha_{nf} = \frac{k_{nf}}{(\rho c_p)_{nf}} = \frac{k_{nf}}{\phi(\rho c_p)_d + (1 - \phi)(\rho c_p)_f}. \tag{8.39}$$

  b. Compare the values of Nusselt number for nanofluids with the values of water for the same Reynolds number. Plot the variation of Nu_{nf}/Nu_{bf} with the volume fractions for all nanofluids.

Chapter 9
Measurement Techniques at Microscales and Nanoscales

In this chapter, the measurement techniques used for measuring the essential properties encountered in microscale and nanoscale energy transport are discussed in detail. In general, measurements play a vital role, followed by theoretical modeling, which always supports the measurements. Therefore, experiments have to be performed even at small scales. However, it is challenging to build small experimental setups for the measurement of temperature, pressure, and flow rate at microscales and nanoscales. Hence, more sophisticated techniques have to be used to probe into smaller scales.

At the microscale, after fabricating a microchannel, its diameter or dimensions have to be calibrated. Assume that the microchannel has a particular surface roughness, which could be of the order of a few microns, and that the diameter and depth of the channel is of the order of a few tens of microns to hundreds of microns.

9.1 Measurement of Dimensions

Therefore, the fundamental first step of measurements is the measurement of the microchannel's dimensions and roughness. This can be done by filling the channel or capillary tube with a denser liquid such as mercury. First, the weight of the empty tube (without mercury) is measured. Next, the tube is filled with mercury and the weight of the tube is measured. The difference between the weights of the tube with mercury and that without mercury is calculated. The difference between the filled tube weight and the empty tube weight divided by the density of mercury gives the total volume of the capillary tube. From this, the diameter of the tube can be estimated.

The diameter measurement also helps measure the surface roughness. If there are small cavities or defects on the surface of the tube, then the liquid will also fill those cavities. The volume of the cavity can be calculated by adopting the aforementioned

© The Author(s) 2025
A. Pattamatta and S. K. Das, *Fundamentals of Nano- and Microscale Heat Transport,*
https://doi.org/10.1007/978-3-031-89613-2_9

procedure. Based on the volume, the surface roughness can be estimated. However, this method is not a perfect tool for measuring the roughness with reasonable accuracy.

Advantages

1. It is a straightforward method.
2. This method enables the accurate determination of the average flowing area of a microchannel along with its length.
3. Only a precision balance is required to estimate the flowing area.

Limitations

Very long microtubes are needed to obtain good accuracy of measurement for very small tubes. The other method for measuring the diameter of microchannels is by passing a liquid through a capillary tube. A pump is not required to fill the tube with mercury and it can be filled manually. From the Poiseuille law, the flow rate can be expressed as a function of pressure drop. Therefore, on measuring the pressure drop using a differential pressure transducer, the mass flow rate can be calculated using the Poiseuille law. The time taken by the liquid to pass through the tube can be measured directly to calculate the mass of the liquid. Therefore, the diameter of the tube can be estimated from the mass since the density of the liquid and the length of the tube are known.

The only difference between this and the previous method lies in the fact that the liquid was static in the previous method. The tube was filled with the liquid and the weight was measured. In the present method, the liquid is pumped into the tube using a pump and the Poiseuille law is used to estimate the mass.

Advantages

1. This technique can be advantageous in reducing the uncertainty in the determination of the inner diameter of a microtube ($\pm 0.2\,\mu$m).
2. The order of accuracy for the average inner diameter measurement for a tube with a nominal diameter of 150 μm is $\pm 0.13\%$.

Limitations

1. For minimal mass flow rates, this technique requires very long time for liquid accumulation.
2. When water is used as a test fluid, care must be taken to estimate the weight of accumulated liquid lost due to evaporation, which strongly depends on the environmental conditions. This aspect can be very critical when lower mass flow rates are generated in smaller microchannels.

Scanning Electron Microscope

A scanning electron microscope is primarily used to determine the nature of the surface roughness and to show the profile of the channel cross section with a high resolution. It can also be used to measure the diameter of capillary tubes or silicon rectangular microchannels bonded with a silicon/glass cover. This can be done by cutting a small portion of the channel cross section and it has to be put under the microscope. The profile of the channel cross section can be determined with a very high resolution. By using an SEM, any tiny blur appearing on the surface in the range of nanometers can be resolved quickly. A schematic diagram of a scanning electron microscope is shown in Fig. 9.1.

Advantages

With magnification and image processing, the dimensions of the channel cross section and the topology of the surface roughness can be determined with reasonable accuracy.

Limitations

When the tested microchannels are closed (e.g., commercial microtubes), this technique applies to the inlet and outlet sections only. No information can be derived on the axial variation of the channel geometry and the roughness distribution along the channel.

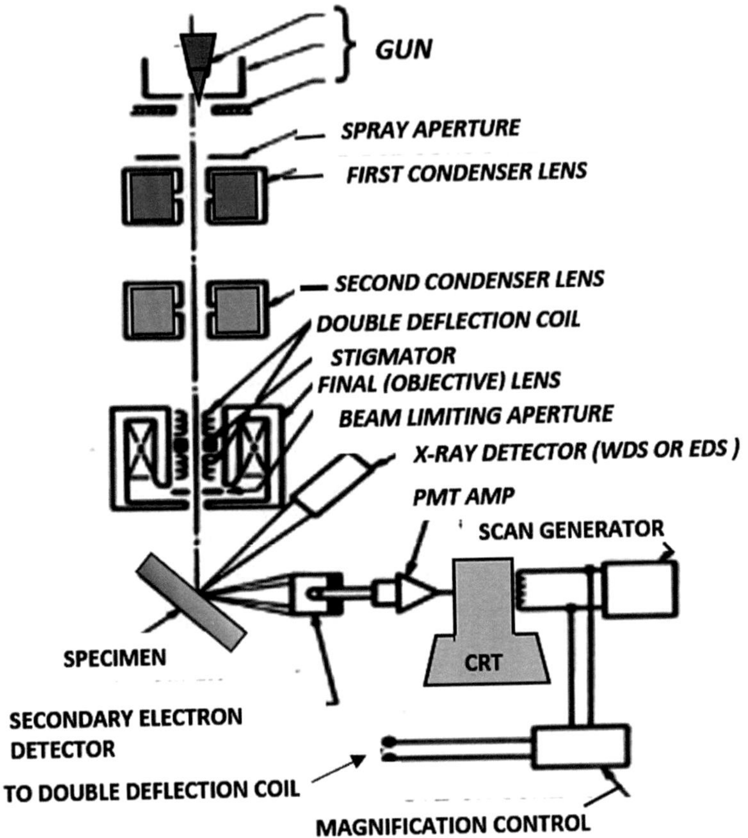

Fig. 9.1 Schematic diagram of a scanning electron microscope

9.2 Measurement of Pressure

The next significant quantity of measurement is pressure. In microchannels, differential pressure transducers are used to measure the pressure drop, which are placed in the reservoirs connected to the microchannel inlet and outlet manifolds. Their presence induces pressure losses at the inlet and outlet of the channel, which can be determined by

$$\Delta p_{\text{in}} = \frac{k_{\text{in}}}{2}\left(\frac{\dot{m}}{\Omega}\right)^2\left(\frac{RT}{p_{\text{in}}}\right),\tag{9.1}$$

$$\Delta p_{\text{out}} = \frac{k_{\text{out}}}{2}\left(\frac{\dot{m}}{\Omega}\right)^2\left(\frac{RT}{p_{\text{out}}}\right).\tag{9.2}$$

In microchannels, the entry and exit losses are very significant compared to macrochannels. These losses mostly depend on the position of the pressure tappings. In general, the tappings are placed at the inlet and outlet manifolds since these are to be positioned at the place where sufficient contact area exists. Therefore, the actual pressure difference includes the contraction and expansion losses at the inlet and outlet, respectively. Hence, the exact value of pressure drop measurement across the microchannel is a complicated process because the pressure tap cannot be placed inside the channel, which disturbs the flow.

Therefore, there arises the need to identify a method to eliminate the contraction and expansion losses. The simplest way is to use the correlations to measure the pressure drop at the inlet and the outlet, which has to be subtracted from the total pressure drop. However, the correlations give only the approximate values. Also, they are not applicable to all kinds of microchannel shapes. Hence, a better way to calculate the pressure drop is by using the tube cutting method.

In this method, the more significant and smaller tubes are taken, and the pressure drop across them is measured. The difference in pressure drop values is calculated so that the common factors such as the inlet and exit losses can be eliminated since the magnitudes of the inlet and exit losses are the same for both long and short tubes. From the difference, the friction factor can be calculated, which gives the pressure drop only within the tube, i.e.,

$$
f = \frac{2\rho}{\mu^2} \left(\frac{\Delta p_{\text{total}}(L_1) - \Delta p_{\text{total}}(L_2)}{L_1 - L_2} \right) \frac{D_h^3}{\text{Re}^2},
\tag{9.3}
$$

where L_1 and L_2 are the lengths of long and short tubes, respectively.

This technique involves repeating the same mass flow rate conditions in both long and short tubes because the pressure drop is a function of mass flow rate. The short and long tubes can be selected from the set of identical tubes. A schematic representation of the measurement principle is shown in Fig. 9.2.

Limitations

1. This method can only be applied to fully developed incompressible flows (liquids or gases under low pressure ratio (P_{in}/P_{out}) where the variation of pressure drop in the tube is linear).
2. This method cannot be applied to compressible and rarefied flows where the pressure distribution along the channel is not linear.
3. This method holds only for low Reynolds numbers, which correspond to low Mach numbers at the outlet for gas flows.

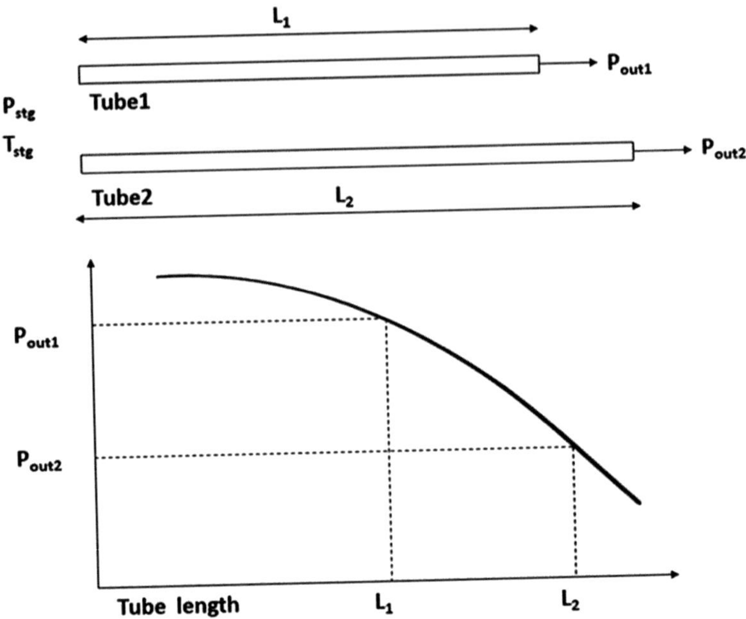

Fig. 9.2 Measurement principle

Pressure-Sensitive Paints

To overcome the limitations of the tube cutting method, the pressure-sensitive paint (PSP) technique was developed. However, this technique requires highly sophisticated equipment compared to the simple differential pressure transducer used in the previous section.

In this technique, a PSP coating is applied on the surface. Then, a normal white light or a laser light is projected onto the coating. The beam emitted by the coating is responsive to the intensity of the projected light and the pressure exerted by the molecules on the surface. If the pressure is on the higher side, the intensity of the emission coming from the paint will also be higher and vice versa. Prior to the experiment, the pressure variation with respect to the intensity of the emitted light should be adequately calibrated to eliminate errors in pressure measurement. A schematic diagram of the pressure-sensitive paint equipment is shown in Fig. 9.3.

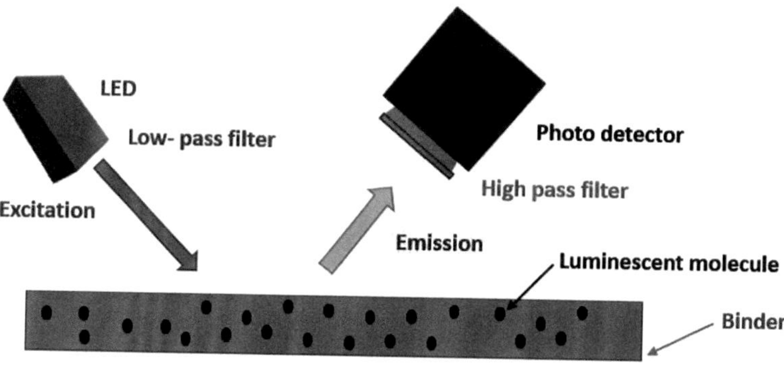

Fig. 9.3 Schematic diagram of pressure-sensitive paint technique

Limitations

1. This technique is based on oxygen quenching. This indicates that the pressure-sensitive paint responds to oxygen molecules rather than to other molecules. Hence, it cannot be applied to test other pure gases such as N_2 and He.
2. The pressure-sensitive paint is coated on a transparent side of the microchannel to ensure the direct contact with oxygen and excitation of luminescent molecules by an external light source. Therefore, the surface should be optically accessible. It is challenging to apply this technique on opaque surfaces.
3. It is impossible to coat pressure-sensitive paints on circular or elliptical micro-tubes.
4. Passages that do not have a transparent side cannot benefit from this optical technique.

Optical Lever Method

The third method for pressure measurement is the optical lever method. This method is more sophisticated than the other two methods and can be applied to a wide range of working fluids. In this method, micrometric pressure tappings are connected to the silicon membrane, which deform according to the local pressure. This means that the deflection is proportional to the pressure exerted onto the membrane. The optical lever equipment is shown in Fig. 9.4.

A sensor can detect the intensity of the laser light reflected or deflected from the silicon membrane. The change in deflection angle can be measured with a photodiode sensor which is precisely moved and positioned. Based on this method, an integrated pressure sensor can be produced. The sensitivity of the pressure sensor can be easily adjusted by changing the spatial resolution. The spatial resolution is the distance between the membrane and the photodiode sensor. Based on the deflected angle, the

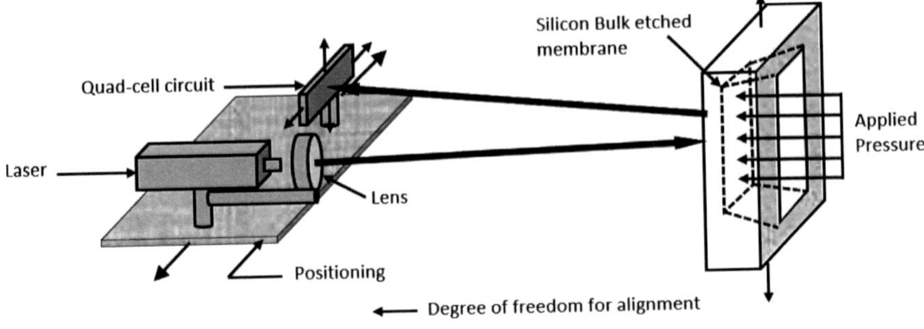

Fig. 9.4 Optical lever method

actual pressure can be estimated. However, proper calibration between the deflected angle and the variation of pressure should be done. The uncertainty in pressure measurement can range from ±2.4% to 13%.

9.3 Measurement of Temperature

The next essential quantities of measurement in microchannels are the temperature and flow measurements.

In conventional thermocouples, a junction of chromium and aluminum produces an emf, which is a function of temperature. The generated emf and the temperature can be related to the Seebeck coefficient. A similar principle can be used at the microscale. However, the difference is that the junction should have a diameter of the order of microns. Otherwise, the thermocouple intrudes into the flow in microchannels. Therefore, it is an intrusive method for measuring the temperature, which is not preferred. The measurement of the bulk fluid temperature along a microchannel cannot be done directly without causing significant disturbance in the fluid flow. Therefore, to avoid disturbances in the fluid flow, microfabrication offers a large variety of devices to achieve local temperature measurements. The existing temperature measurement devices can be classified into three groups:

1. a thin-film microresistance temperature detector (μRTD),
2. thin-film thermocouples (μTFTC),
3. semiconducting sensors (SC).

Microresistance Temperature Detector (μRTD)

The functioning principle of a μRTD is the same as that of the classical RTD in which the temperature-dependent electrical resistivity of a material is used for estimating the temperature. Proper calibration between the resistance and temperature must be done. Calibration is not required sometimes because the value of dT/dr is constant and is given over a certain temperature range by the manufacturer itself. Therefore, the temperature value can be obtained directly from the temperature coefficient of electrical resistance.

This method is one of the simplest ways of measuring the temperature since only a metallic element with a very high temperature coefficient of electrical resistance is used. Platinum is commonly used in different applications for its catalytic property in combination with some gases or, at other times, its inert property in combination with a large variety of fluids.

Advantages

1. This device is straightforward to fabricate.
2. It has a simple linear response. Therefore, the same RTD used for the macroscale can also be miniaturized.

Thin-Film Thermocouple (μTFTC)

Thin-film thermocouples are the active elements that use the Seebeck effect to measure the temperature of a two-metal junction. The microfabrication of μTFTCs is complex depending on the existing material constraints on the specific microdevice needed. Zhang et al. proposed a Chromel thin-film thermocouple embedded on a Ni substrate and demonstrated that its behavior is similar to that of standard K-type thermocouples.

Advantage

KμTFTCs with a junction of 25 μm and a film thickness of 150 nm are able to provide a good sensitivity (40.4 μV $°C^{-1}$) up to 800 °C and a fast response time (28 ns).

Limitation

It is difficult to fabricate and maintain the surface quality of this kind of microcomponent.

Semiconducting Sensors

Semiconducting sensors are based on semiconducting devices rather than on metallic elements, where the junction voltage is proportional to the absolute junction temperature. Hence, in the case of a p–n junction diode, the voltage measured across the p–n junction is a direct function of the temperature. Therefore, proper calibration between the voltage and temperature can be done to measure the actual temperatures.

At room temperature, silicon p–n junctions have a forward voltage drop of 0.7 V, and this voltage decreases by 2 mV for every degree of increase. The typical temperature range of these sensors could be 55–1750 °C with a typical accuracy between ± 0.1 and ± 0.30 °C.

Advantage

Their flexibility and high scale of integration enable us to expand a single sensor to an array of sensors.

Limitation

Leakage currents may exist at elevated temperatures. The other vital component is the measurement of velocity. For the determination of the local velocity of a gas in microchannels, three optical velocimetry techniques have been proposed in the literature as follows:

1. laser Doppler anemometry (LDA) (used for >1 mm channels),
2. microparticle image velocimetry (μPIV),
3. molecular-tagging velocimetry (MT-V).

Laser Doppler anemometry is used for larger channel dimensions, and μPIV is used for small channel dimensions. The conventional PIV has to be scaled down to capture the micron-size resolution. Molecular tagging velocimetry is an expensive method.

Microparticle Image Velocimetry (μPIV)

PIV is a nonintrusive measurement technique. The equipment consists of a laser source and a high-speed camera. From one end to the other end, the system has to be captured by the high-speed camera. Therefore, it should have an optical access for the laser to penetrate into the system. In addition, it should have an optical access for the camera to capture the light scattering and coming out of the system. Therefore, it should have a two-way optical access. Moreover, it should also contain seeding particles of negligible density compared to the actual fluid for which the velocity has to be measured. If the density of the seeding particles is high, then an accurate velocity field cannot be obtained because the seeding particles will have their own momentum, which is different from that of the parent fluid. Therefore, the particles should be quite light as well as luminescent. This indicates that while radiating them with a laser of a particular wavelength, they should start glowing so that they can be captured by the high-speed camera.

Several millions of tracer particles are added to the main fluid element of the system. Then, the laser is pulsed every few nanoseconds or microseconds. One can keep track of the particles with the laser glow. The images can be recorded with a CCD camera at different time intervals every few microseconds or nanoseconds. This can be done by looking at the snapshots at two different positions after times t and $t + \delta t$, which are x_t and $x_t + \delta t$, respectively. The difference between the positions of the particle divided by the time interval gives the magnitude of the velocity. For every particle, the corresponding velocity vectors can be calculated, i.e.,

$$u \approx \frac{x_{t+\Delta t} - x_t}{\Delta t}. \tag{9.4}$$

Therefore, the local velocity vectors all over the entire system can be obtained from the 3-D snapshots. This indicates that the system should have two cameras. By taking two different planes, one can construct a 3-D motion of these particles, which gives the complete 3-D velocity field. In the case of μPIV, the system is much smaller, and the response time has to be very quick. The laser pulse should be small. A high-speed camera is used to take the images at very high frames per second. The common differences between μPIV and PIV are tabulated in Table 9.1.

The interrogation cell is the particular unit cell across which a particular tracer can be tracked within δt. The tracer can move from one point to another within the interrogation cell over δt. If the interrogation cell width is too small, then the tracer comes out of the cell. If δt is too large, the tracer within the particular interrogation cell cannot be captured. Therefore, the interrogation cell size should be very small to

Table 9.1 Difference between μPIV and PIV

Aspect	Micro-PIV	PIV
Field of view	50–500 μ	30–300 mm
Vector spacing	1–10 μm	1–10 mm
	Depth of the field of the microscope 1 μm	Laser sheet thickness 1 mm
	Small enough to follow flow, does not clog the device and does not alter the fluid property	Small enough to track flow, needs to be detectable by the camera
Particle diameter (D_p)	0.3–0.7 μm	3–30 μm

resolve the local velocity data more accurately. Hence, the particle diameter has to be even smaller than the interrogation cell size. Therefore, in the case of μPIV, if the interrogation cell size is 2–20 mμ, then the particle diameter should be of the order of 0.3–0.7 mμ. In the conventional PIV, the particle size is of the order of 3–30 μm. The laser sheet thickness should also be low and should be of the order of microns, whereas in the conventional PIV, the laser sheet thickness should be of the order of millimeters.

In the case of μPIV, the laser goes through a beam of expanding optics, which converts the point source into a sheet of light. This sheet is passed through a filter cube, which changes the angle of this beam by 90°. Then the beam enters the microfluidic device located at the bottom. The sheet of light interacts with the seeded particles inside the device. There is a CCD camera placed right at the top to take snapshots of the seeded particles at regular time intervals.

Advantages

1. It is a noninvasive global flow measurement technique.
2. The systems provide accurate measurement of the fluid flow in micro-fluidics.
3. It utilizes special concepts such as illumination, scattered light collection, seed particle concentration, optical design, and data analysis to offer the most powerful and versatile system.

Limitations

1. It reduces flare from the walls by using fluorescent particles and filters the illuminating wavelength.
2. The particles must be very small to fit in the small measurement volume because small particles have significant Brownian motion.
3. Volume illumination, instead of a light sheet, creates unwanted images from out-of-focus particles.
4. It reduces background glow, which implies shallow channel flow and low-concentration particles.
5. Short times are noted between pulses and no pulses, even in slow flow; it needs pulsed lasers.

Besides μPIV, molecular tagging velocimetry is used to measure the local velocity. This method is similar to PIV, where the velocity is obtained from the displacement of tracer particles. But in molecular tagging velocimetry, the molecules of the working fluid itself respond to the incoming laser pulse. This indicates that the pulse laser should tag the region over which the velocity field has to be calculated. Then the imaging laser will send a sheet beam of light through the tagged region so that the molecules change their color responding to the incident laser. This phenomenon should be used as a signature to track the velocity vectors of the particular molecules.

Therefore, the molecules of the working fluid themselves act as elements to trace the fluid velocities. The advantage is that the use of conventional seeding particles is not needed in this method. A major limitation of this method is that it can be used only for liquid flows because most of the liquid molecules are more responsive to lasers, whereas gas molecules are not.

9.4 Measurement of Flow Rate

To measure the flow rate, a flow meter has to be connected at the inlet or exit to measure the actual flow rate. This is challenging because the flow rates are of the order of less than 10^{-8} kg/s. Therefore, there are different methods such as direct and indirect methods to check the mass flow rate.

The volumetric flow meters give an indication of the flow rate regardless of the fluid tested. The mass flow meters can be used only with the fluid for which they have been calibrated. An indirect way of measurement is by checking the values taken by other measurable quantities such as pressure, forces, weight, volume, and temperature.

Sometimes, the flow rate in a small beaker over a period of time is collected and then the difference in the volume that is actually flowing is checked. It is not very uncommon for researchers to use very simple methods, such as collecting flow rate in a beaker over a period of time to calculate mass flow rates rather than using expensive mass flow meters.

9.5 Measurement of Thermal Conductivity

The pressure, temperature, flow rate, and velocity measurements are related to microchannels. In the nanoscale, the thermophysical properties such as thermal conductivity and heat capacity are required to solve the Boltzmann transport equation. Thermal conductivity is a property of a material, which is a measure of its ability to conduct heat. It can be represented by Fourier's law as follows:

$$\vec{q} = -k \vec{\nabla} T. \tag{9.5}$$

One of the most common ways of measuring thermal conductivity at conventional scales is the guarded hot-plate technique. It is a steady-state measurement technique, where the material is simply sandwiched between two metallic junctions, in which one junction is at a high temperature and the other is at a low temperature. By measuring the temperature difference and heat flux, the effective thermal conductivity can be calculated. On the one hand, the applied heat flux magnitude is known. On the other hand, a constant temperature has to be maintained. On measuring the hot-side temperature, the cold-side temperature, and the heat flux, thermal conductivity can be estimated using Fourier's law. In the case of nanofilms, temperatures cannot be measured directly. Therefore, measuring the thermal conductivity without measuring the temperature is really challenging. To overcome this difficulty, David Khail proposed the most common method in the 1990s called the 3-omega method. The 3-omega method is a very novel, innovative technique for measuring thermal conductivity of bulk materials or nanoscale materials without measuring the temperature.

The thermal conductivity measurement techniques can be classified into two: the measurement at steady state and transient. The steady-state method includes the guarded hot-plate method and the radial hot-plate method. The transient method includes the 3-omega method and the transient hot-wire method. In the case of the transient hot-wire method, a probe is immersed in the fluid and the response period is noted. Based on the change in the temperature of the probe, the thermal conductivity of the liquid can be calculated.

In the case of nanoscale films, this method cannot be used directly. Therefore, the 3-omega technique is used. The 3-omega technique falls into the transient measurement category. It is not a conventional steady-state method. In this method, the samples for which the thermal conductivity has to be measured are etched and a gold film is deposited on them in such a way that a four-point probe is deposited on them. This four-point probe will have four contact pads. Across two contact pads, the input current I is supplied at a frequency. The voltage is measured across the other two contact pads. Therefore, the same probe acts as both a heater and a sensor. An increase in temperature can be observed because of the resistance across the substrate, which causes a change in the temperature over a period of time, which, in turn, induces a voltage across the other two contact parts. Therefore, on supplying a current of frequency 1 omega,

$$I \sim 1\omega.$$

The temperature response varies according to I square, which is of the order of two times omega:

$$T \propto H \propto I^2 \sim 2\omega,$$

and the resistance also changes as a function of temperature, which is two times omega:

$$R \propto T \sim 2\omega.$$

The product of I and R has two components, which are of the order of three omega:

$$V \sim IR \sim 3\omega,$$

where $I = I_0 \cos \omega t$.

The specialty of the 3-omega component is that extracting the 3-omega component directly gives the increase in temperature at the surface of the material so that the thermal conductivity can be calculated with this particular value. Therefore, taking the product of I and R, two components come from omega and three omega. The amplitude of the three-omega component is called $v_{3\omega}$. This is the voltage corresponding to the third harmonic or 3-omega, i.e.,

$$V = I_0 \cos(\omega t) R_0 (1 + TCR.\Delta T(0) \cos(2\omega t)),\tag{9.6}$$

$$V = I_0 \cos(\omega t) + \frac{1}{2} I_0 R_0.TCR.\Delta T(0)(\cos(3\omega t) + \cos(\omega t)).\tag{9.7}$$

Therefore, the coefficient of resistance to temperature can be defined as

$$k = \frac{V^3 \ln \frac{\omega_1}{\omega_2}}{4\pi l R^2 (V_{3\omega,1} - V_{3\omega,2})} \frac{dR}{dT}.\tag{9.8}$$

Based on this, the thermal conductivity of the bulk film can be expressed directly. Therefore, this method is used for a bulk film of the order of micron size.

In the nanoscale, a nanothin film has to be deposited onto the top of the bulk material, which is electrically insulating. The sample used for measuring the thermal conductivity should be an electrical insulator. Otherwise, apart from heat conduction, there is also a possibility of electrical conduction through the film, which may corrupt the data. Therefore, for thin films, the total increase in temperature for both the film and the substrate is measured and the contribution of the substrate is subtracted. First, the experiment is done for the substrate and the increase in temperature for the substrate is measured. Then, the experimentation is done for both the film and the substrate to measure the total increase in temperature. The difference between these two values gives the increase in temperature only due to the thin film from which the thermal conductivity can be calculated as follows:

$$\Delta T_{tot}(0) = 2C \frac{dT}{dR} \frac{R_0}{V_{1\omega}},\tag{9.9}$$

$$\Delta T_{sub}(0) = \frac{p}{\pi l k_{sub}} \left[\frac{1}{2} \ln \left(\frac{k_{sub}}{P_{sub}\left(\frac{w}{2}\right)^2} \right) + \eta_{sub} - \frac{1}{2} \ln(2\omega) \right],\tag{9.10}$$

$$k_{film} = \frac{pt}{wl \Delta t_{film}}.\tag{9.11}$$

This method can be extended to the calculation of the thermal interface resistance. In the case of a combination of a metal and a semiconductor, there is a temperature drop at the interface of these two different materials.

Table 9.2 Comparison of the results obtained by Srinikaeth and Pattamatta (2016) with those of Chien et al. [18]

Sample	Srinikaeth and Pattamatta (2016)	Chien et al. [18]
Si sandwiched	3.473 E-08	2.44 E-08
Cr sandwiched	2.158 E-08	3.7 E-09

For the thermal interface resistance measurement,

$$\Delta T_{\text{total},1} = \Delta T_{\text{SiO2}} + \Delta T_{\text{cr}} + \Delta T_{\text{SiO2}} + \Delta T_{\text{Si}} + 2\Delta T_{\text{Cr}-\text{SiO2}},$$

$$\Delta T_{\text{total},2} = 2\Delta T_{\text{SiO2}} + \Delta T_{\text{Si}},$$

$$\Delta T_{\text{total},1} - \Delta T_{\text{total},2} = \Delta T_{\text{cr}} + 2\Delta T_{\text{Cr}-\text{SiO2}}, \tag{9.12}$$

$$\Delta T_{\text{Cr}-\text{SiO2}} \approx \frac{\Delta T_{\text{total},1} - \Delta T_{\text{total},2}}{2}.$$

Therefore, the thermal interface resistance can be calculated as

$$R_{th} = \frac{\Delta T_{\text{interface}}}{q_{\text{th}}}. \tag{9.13}$$

The thermal interface resistance was calculated experimentally by Sriniketh and Pattamatta, and the results were compared with those of Chien et al. [18], as shown in Table 9.2, for the chromium sandwich between two silicon dioxide substrates and also for a silicon sandwich between two silicon dioxide substrates.

Figures 9.5 and 9.6 show the variation of thermal conductivity as a function of film thickness and temperature, respectively. From Fig. 9.5, one can observe that as the size of the film decreases, the thermal conductivity decreases. Two different values of thermal conductivity can be observed at a thickness of 1000 nm because the method of preparation is different. But at nanoscales, they both converged. Similarly, samples of different sizes, such as 100 nm and 500 nm, exhibit a strong functional dependence on temperature, as shown in Fig. 9.6.

9.6 Uncertainty Analysis

After measuring all the quantities, a very important analysis called uncertainty analysis should be carried out to check the accuracy of the experimental values. Experiments are never 100% accurate because they all involve uncertainty in every stage. For example, to calculate the friction factor from experiments, other quantities such as

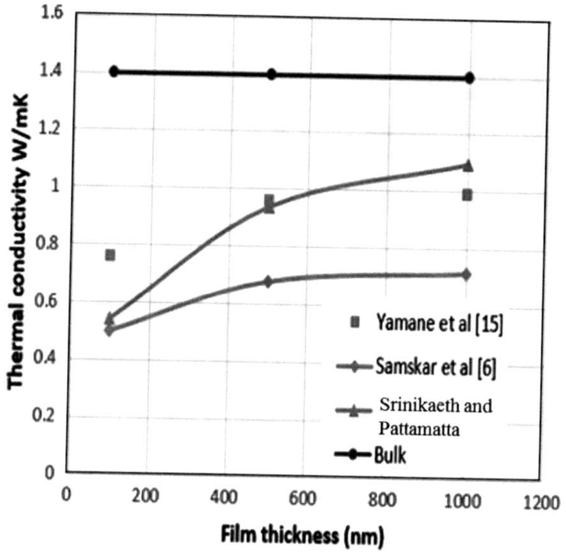

Fig. 9.5 Variation of thermal conductivity with film thickness [89]

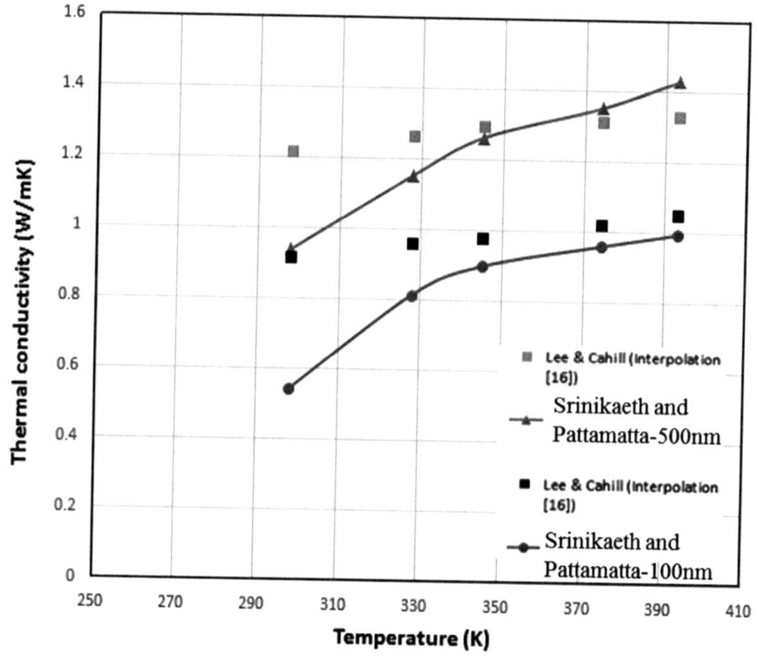

Fig. 9.6 Variation of thermal conductivity with temperature [89]

pressure drop, velocity, density, dimension length, and diameter need to be measured. The uncertainty is more while calculating the dimensions at micron size.

Therefore, one should know the uncertainty in measuring parameters, such as length, diameter, pressure drop, density, and velocity. Therefore, the friction factor is a general function of

$$f = f(x_1, x_2, \ldots, x_n). \tag{9.14}$$

Independently, measurements have to be taken and uncertainty analysis has to be carried out again to calculate the global uncertainty in friction factor. The absolute uncertainty in friction factor can be defined as the square root of the sum of the squares of all the individual uncertainties in the parameters multiplied by the derivative of the friction factor with respect to individual parameters, such as length, diameter, density, and pressure drop. The absolute uncertainty in friction factor is given by

$$\delta f = \sqrt{\sum_{i=1}^{n} \left(\frac{\partial f}{\partial x_i}\right)^2 (\delta x_i)^2 + 2 \sum_{i=1}^{n} \sum_{j=i+1}^{n} \frac{\partial f}{\partial x_i} \frac{\partial f}{\partial x_j} \operatorname{cov}(x_i, x_j)}. \tag{9.15}$$

On giving weightage to certain parameters depending on the importance of those parameters in calculating the friction factor, the uncertainty in friction factor can be calculated as

$$\frac{\delta f}{f} = \sqrt{k_1^2 \left(\frac{\delta D_h}{D_h}\right)^2 + k_2^2 \left(\frac{\delta l}{l}\right)^2 + k_3^2 \left(\frac{\delta T}{T}\right)^2 + k_4^2 \left(\frac{\Delta \delta p}{p}\right)^2 + k_5^2 \left(\frac{\delta p_{out}}{p_{out}}\right)^2 + k_6^2 \left(\frac{\delta \dot{m}}{\dot{m}}\right)^2}. \tag{9.16}$$

This is called the relative uncertainty. The weightage given to each parameter is called the sensitivity coefficient. The relative uncertainty gives the % uncertainty in determining the friction factors. It could be plus or minus 5–10%. This is the standard procedure used across all the measurement analyses.

In this chapter, various measurement techniques used for measuring dimensions, pressure, temperature, flow rate, thermal conductivity are elucidated in detail. Finally, an uncertainty analysis is illustrated to check the accuracy of the experimental values.

Exercise Problem

1. Uncertainty analysis for a bulk SiO_2 film in 3ω method is as follows. Find the uncertainty in the measurement of thermal conductivity of bulk SiO_2 film.

Parameter	Nominal value (u)	Uncertainty (du)	% change (du/u $\times$ 100)
Length	1000 m	5 m	0.5
Resistance	9.19	0.1	1.088
TCR	0.0014 K−1	0.0001 K−1	7.143
Voltage	0.6 V	0.1 mV	0.0167
V3	0.298 mV	0.1 V	0.0336
Thermal conductivity	0.962 W/mK		

Chapter 10
Numerical Simulations of Nanoscale Heat Transport

Heat transfer at the nanoscale can be categorized into distinct regimes, illustrated in Fig. 10.1. When the structural dimensions are approximately an order of magnitude larger than the mean free path, the heat transport is diffusive. As the size of the structure approaches or falls below the mean free path, size-related effects emerge, eventually leading to ballistic transport. These size effects are further classified into the classical size effect regime and the quantum size effect regime, depending on the ratio of the characteristic length to the carrier wavelength. These regimes are also known as the particle and wave regimes, or incoherent and coherent transport regimes. In the classical size effect regimes, the phase of the energy carriers can be disregarded, and their trajectories can be tracked instead. Conversely, in the quantum size effect regime, the phase information becomes crucial and must be taken into account. The determination of whether the transport falls into the classical or quantum size effect regime is contingent on whether the interface scattering processes can preserve the phases of the waves, a factor influenced by the ratio of interface roughness to the carrier wavelength, as depicted in Fig. 10.2.

10.1 Governing Equations and Formulation

10.1.1 Boltzmann Transport Equation

The Boltzmann Transport Equation is applicable to various energy carriers, including dilute gas molecules, electrons, photons, and phonons. It is highly versatile, as it can encompass the macroscopic transport behavior of particles. Laws such as Fourier's Law, Ohm's Law, Newton's shear stress law, Fick's Law, and the hyperbolic heat equation can all be derived from the Boltzmann equation in the macroscale limit, employing suitable approximations. Furthermore, the Boltzmann equation allows the derivation of well-known conservation equations for mass, momentum, energy,

A. Pattamatta and S. K. Das, *Fundamentals of Nano- and Microscale Heat Transport*, https://doi.org/10.1007/978-3-031-89613-2_10

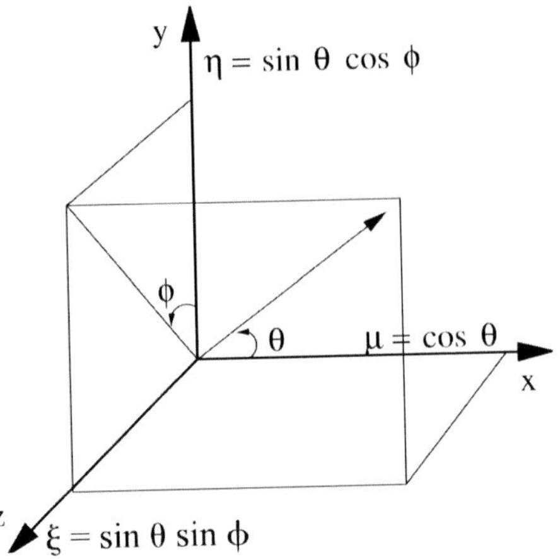

Fig. 10.1 Coordinate system used in the simulations

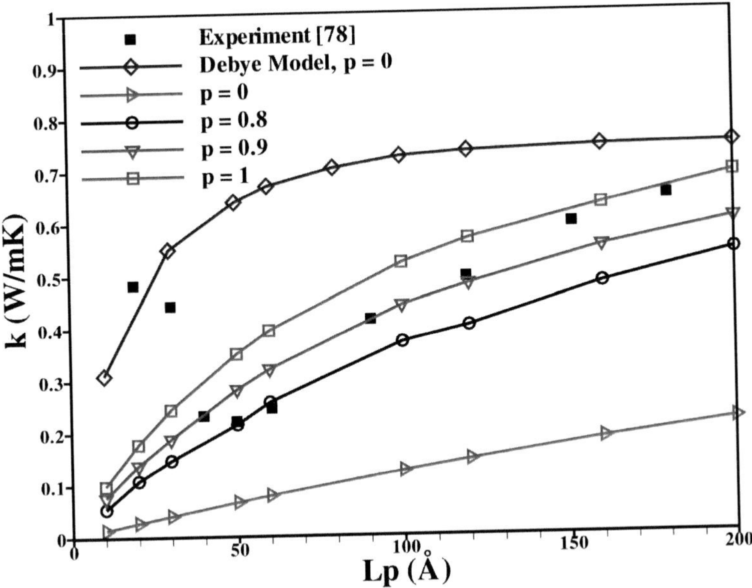

Fig. 10.2 Variation of thermal conductivity with period thickness for $Bi_2Te_3 - Sb_2Te_3$ superlattice

and electro-hydrodynamics. In its general form, the Boltzmann Transport Equation can be expressed as

$$\frac{\partial f}{\partial t} + \mathbf{v}.\nabla_{\vec{r}} f + \vec{F}.\nabla_{\vec{p}} f = \left(\frac{\partial f}{\partial t}\right)_c \tag{10.1}$$

where f is the statistical distribution function of carriers, which depends on time t, position vector $\vec{r}$, and momentum vector $\vec{p}$. Here $\overrightarrow{F}$ is the force applied to the particles. The terms on the left hand side of Eq. (10.1) represent the time rate of change of the distribution function, convective transport of the function due to the carrier group velocity, and acceleration due to external forces, respectively. The term on the right hand side of Eq. (10.1) represents the scattering term, which restores the system to equilibrium. Quantum mechanical principles are often used to deal with scattering. A general expression of the scattering integral can be formally written, based on the scattering probability and the distribution function. The most basic scattering mechanism is due to the anharmonicity of the lattice potential, which permits phonon–phonon scattering, including normal phonon–phonon scattering and Umklapp phonon–phonon scattering. In nanostructures, phonon-interface scattering dominates the transport at room and higher temperatures. This scattering mechanism is dominant at very low temperature in bulk materials. The evaluation of scattering integrals for phonons is very difficult. Generally, the scattering term can be written as

$$\left(\frac{\partial f}{\partial t}\right)_c = \left(\frac{\partial f}{\partial t}\right)_{e-p} + \left(\frac{\partial f}{\partial t}\right)_{p-p} \tag{10.2}$$

where $(\partial f/\partial t)_{e-p}$ is the net phonon generation rate from electron–phonon interaction. $(\partial f/\partial t)_{p-p}$ denotes the phonon number change due to the phonon–phonon scattering, phonon scattering by lattice defects, and phonon-boundary scattering. Often the relaxation time approximation is used

$$\left(\frac{\partial f}{\partial t}\right)_c = \frac{f_0 - f}{\tau} \tag{10.3}$$

where f_0 is the Bose–Einstein distribution for phonons at equilibrium temperature T, and τ is the relaxation time as a function of temperature and frequency.

When estimating the effect of several types of scattering centers, each related to a specific relaxation time τ_i, the total time between independent scattering events can be estimated from Mathiessen's rule, since the collision term in the Boltzmann equation is proportional to the collision rate and is additive, as shown

$$\frac{1}{\tau} = \sum_i \frac{1}{\tau_i}. \tag{10.4}$$

In analogy to the equation of radiative transfer for photons, the equation of phonon radiative transfer has been derived as

$$\frac{\partial I_\omega}{\partial t} + \mathbf{v}.\nabla_r I_\omega = \frac{I_{\omega,0} - I_\omega}{\tau_\omega} + S_\omega \tag{10.5}$$

where τ_ω is the phonon relaxation time, I_ω is the phonon intensity, and S_ω is due to the phonon generation through the scattering with other carriers, for example, electron–phonon scattering and photon–phonon scattering. The phonon intensity is defined as

$$I_\omega = v_\omega \hbar \omega f D(\omega)/4\pi \tag{10.6}$$

where $D(\omega)$ is the phonon density of states per unit volume, which should be obtained from the phonon dispersion relation of the specific material.

A rigorous phonon transport simulation should incorporate the frequency dependence of the phonon relaxation time and group velocity, and thus should account for interactions among the dispersive phonons of different frequencies. However, this requires solution of the phonon BTE for many different frequencies. The existing theories for the frequency dependence of the relaxation time contain large uncertainties because they are based on many approximations and rely on the fitting parameters from experimental data. Generally, an average relaxation time is a good approximation for thermal conductivity modeling across interfaces. Therefore, we will use a frequency independent phonon relaxation time and phonon group velocity for simplicity, i.e., a phonon gray medium approximation.

For a 3-D system, using the coordinate system shown in Fig. 10.1, the frequency independent BTE can be expressed as

$$\frac{\partial I}{\partial t} + v\left(\mu\frac{\partial I}{\partial x} + \eta\frac{\partial I}{\partial y} + \xi\frac{\partial I}{\partial z}\right) = \frac{I_{eq} - I}{\tau} \quad \text{where } I_{eq} = \frac{1}{4\pi}\int_0^{2\pi}\int_{-1}^{1} I\,d\mu\,d\phi \tag{10.7}$$

where μ, η, and ξ are the x-, y-, and z-direction cosines respectively, and v is the average phonon group velocity.

10.1.2 Dispersion Method for Calculating Phonon Properties

The material properties used in the computation are evaluated using the sine function phonon dispersion model. This model uses the frequency averaged specific heat and velocity to calculate the phonon mean free path from the relation

$$k = \frac{\Lambda}{3}\sum_p \int_0^{\omega_{mp}} C_{\omega p}(\omega)\,v_p(\omega)\,d\omega \tag{10.8}$$

where ω_{mp} is the maximum allowable frequency corresponding to the polarization p and is related to the Debye temperature. The phonon dispersion relations for nanostructures are obtained along a particular crystal direction and are assumed isotropic in other directions. They are approximated as sine function similar to that of a linear atomic chain.

$$\omega_p = \omega_{mp}\,\sin(\vec{k}a/2) \tag{10.9}$$

where a is the equivalent atomic separation of an isotropic medium which can be determined from

$$a = \pi \left(6\pi^2/V\right)^{-\frac{1}{3}}.$$

Here, V is the volume of the nanostructure molecule. The group velocity and the specific heat for each polarization mode can be expressed as

$$v_p = \frac{\omega_{mp}a}{2} \cos\left(\frac{\vec{k}a}{2}\right) \tag{10.10}$$

$$C_{\omega p} = \frac{4\hbar^2}{\pi^2 k_b a^3 T^2 \omega_{mp}} \frac{\left[\sin^{-1}\left(\omega/\omega_{mp}\right)\right]^2}{\cos\left(\vec{k}a/2\right)} \frac{\omega^2 \exp\left(\hbar\omega/k_b T\right)}{\left[\exp\left(\hbar\omega/k_b T\right) - 1\right]^2}. \tag{10.11}$$

In the calculation of phonon group velocities, only the acoustic phonon contribution is accounted while neglecting the optical phonon contribution, since, the slopes of the optical phonon dispersion curves are small. The expression for the total specific heat is obtained by integrating the frequency dependant specific heat over the entire frequency range and summing over the acoustic polarization modes

$$C = \sum_p \int_0^{\omega_{mp}} C_{\omega p} \, d\omega. \tag{10.12}$$

The mean free path can also be calculated from Eq. (10.8). The average group velocity of the phonons are calculated from the expression $k = \frac{1}{3} C v \Lambda$, using the bulk thermal conductivity, mean free path, and the total specific heat calculated from Eqs. (10.8) and (10.12).

The Debye model assumes a linear dispersion relationship between the phonon frequency and the wave vector. However, the phonon dispersion model takes into account the nonlinear phonon dispersion and therefore gives a more accurate estimate of the material properties.

10.1.3 Interface Treatment

The BTE for the phonon intensity is solved in conjunction with suitable boundary and interface treatment. The treatment of interfaces between the two materials significantly affects the thermal characteristics of the nanostructure. The interfaces between the two materials are modeled as either totally diffuse or totally specular or as a combination of both. For totally diffuse interfaces, a model called the Diffuse Mismatch Model (DMM) is used. This model makes an assumption that the phonons emerging from an interface are independent of the phonons incident on the interface. The specular scattering off an interface is modeled using the elastic Acoustic Mismatch Model (AMM).

The intensity of the phonons using the DMM for diffuse interfaces and AMM for specular interfaces are separately calculated and are combined using the interface specularity parameter p as

$$p = \exp\left(-\frac{16\pi^3\delta^2}{\lambda^2}\right), \tag{10.13}$$

where δ is the interface roughness and λ is the characteristic phonon wavelength. If the specularity parameter p is zero, it indicates a totally diffuse or a rough interface and a value of p equal to one indicates a specular or a smooth interface. A real interface is neither truly diffuse nor specular and hence cannot be accurately represented by the two limits. The interfaces can be modeled as diffuse if the interface roughness is greater than the phonon wavelength. Even for values of interface roughness less than the phonon wavelength, Eq. (10.13) predicts a high probability for diffuse scattering. Therefore, the value of p is explicitly chosen to combine the diffuse and specular interface treatment models.

10.1.4   Solution Methodology

Modeling heat conduction processes in devices that encompass a range of length and time scales, from nanoscale to macroscale, poses significant challenges. The complexity of such systems requires higher order schemes to accurately capture the diverse length and time scales involved. The Runge–Kutta Discontinuous Galerkin (RKDG) method is employed for the numerical solution of the Boltzmann Transport Equation (BTE), utilizing higher order polynomials for spatial and angular resolution. The RKDG method combines the local conservation property of finite volume schemes with the higher accuracy of finite element methods. This method proves advantageous for problems described by hyperbolic equations, such as those encountered in fluid flows or thermal radiation, offering benefits like higher order local approximation using polynomials, ease of parallelization, element-wise conservation, and geometric flexibility.

Temporal discretization with high resolution is achieved through a four-stage Runge–Kutta (RK) time-stepping scheme, ensuring numerical stability. Spatial discretization utilizes both piecewise constant and linear Legendre basis functions. Gaussian quadrature points and weights are employed for spatial integrations, and the flux at interfaces is approximated using a monotone "upwind" numerical flux scheme.

The computational demands for the numerical solution of the BTE in nanocomposites are prohibitive on a single processor due to the extensive grid size required for detailed spatial and angular resolution. For nanoparticles, a grid size of $128 \times 128 \times 128 \times 30 \times 20$, encompassing three spatial and two angular dimensions, is used.

Solving the 3-D BTE on a single processor would take approximately 3 months for the former case and around 20 days for the latter. Consequently, a parallel algorithm is developed to accelerate computations, parallelizing the solver in the x- and y-directions using Message Passing Interface (MPI) communication routines. The computational domain is divided into subdomains, each assigned to a processor, with the efficiency of the parallel code influenced by the ratio between communication time and total computation time. Scalability is assessed by performing simulations with a fixed subdomain size on an increasing number of processors.

10.2 Numerical Solution to Heat Conduction in Thin Films and Superlattice

The validation of the Boltzmann Transport Equation (BTE) is carried out for steady-state heat transfer in the $Bi_2Te_3 - Sb_2Te_3$ superlattice. This superlattice comprises regularly repeating stacks of Bi_2Te_3 and Sb_2Te_3 layers, facilitating heat transport across these layers. It proves to be a valuable tool for examining the impact of interfaces on reducing thermal conductivity. Figure 10.2 illustrates the effective thermal conductivity, denoted as k, plotted against the thickness of the superlattice period. The graph reveals a decrease in thermal conductivity as the period thickness of the superlattice decreases. In this context, the period thickness refers to the combined thicknesses of the two materials. This reduction is attributed to thermal boundary resistance impeding phonon transport when the mean free path (MFP) of phonons exceeds half the period thickness.

An investigation into the influence of the specularity parameter, p, is conducted to comprehend the effect of interface roughness on the thermal conductivity of the superlattice. The study indicates an augmentation in thermal conductivity with increasing values of p. In scenarios of diffuse interfaces, phonon scattering occurs in all directions, leading to diminished heat flow and consequently a lower value of k. On the contrary, when $p = 1$, a majority of scattering occurs along the heat flow direction, resulting in a higher value of k.

The experimental findings from Venkatasubramanian [97] are also depicted in Fig. 10.2. These results exhibit a similar decline in thermal conductivity as the period thickness decreases, reaching a minimum around 50 Å. However, for thickness values below 50 Å, there is an observed increase in thermal conductivity, approaching the bulk value. While the precise reasons for this behavior remain unclear, some explanations grounded in the wave nature of phonons have been proposed in the literature.

Despite the experimental data exhibiting variability within the 0.8–1 range, the predicted thermal conductivity for $p = 0.9$ yields the smallest root mean square (RMS) error at 15% when compared to the data. In contrast, $p = 0$ shows the maximum error at 76%. This suggests that actual interface scattering cannot be accurately modeled as entirely diffuse or entirely specular. The Debye model's predictions for thermal conductivity do not align with the experimental data. This discrepancy arises from the mean free path (MFP) of phonons calculated using the Debye model being smaller than the superlattice thickness, even for a period thickness of 50 Å. Consequently, the results from the Debye model are closer to the macroscopic regime and fail to capture the ballistic effects observed in the experiment.

10.3 Numerical Solution to Heat Transport in Nanocomposites

Nanocomposites have the potential to reduce thermal conductivity, presenting an avenue to extrapolate nanoscale effects observed in superlattices to practical thermoelectric devices. Despite their significance in applications like thermoelectric devices, there is a limited number of theoretical and computational studies on the thermal conductivity of nanocomposites. This section employs the deterministic solution of the phonon Boltzmann equation to investigate the thermal conductivity of 3-D nanocomposites, wherein nanoparticles are periodically arranged in a host matrix. The outcomes of this study can guide the development of highly efficient thermoelectric materials and thermal interface materials for micro/nano electronic applications requiring high thermal conductivity.

The schematic of the chosen nanoparticle composite for the simulations and the 3-D coordinate system used are depicted in Fig. 10.3. The nanoparticle composite comprises cubic or non-cubic nanoparticles dispersed in a cubic host, as illustrated in Fig. 10.3a. Although, in practical scenarios, these nanoparticles may be randomly distributed within the host, the present study assumes a uniform distribution for computational simplicity. The host materials considered are Bismuth Telluride (Bi_2Te_3), Germanium (Ge), and Gallium Arsenide ($GaAs$), while the corresponding particle materials are Antimony Telluride (Sb_2Te_3), Silicon (Si), and Aluminum Arsenide ($AlAs$). The nanoparticles are assumed to be uniformly distributed within the host material. Heat is applied along the x- and z-directions. The 3-D unit cell shown in

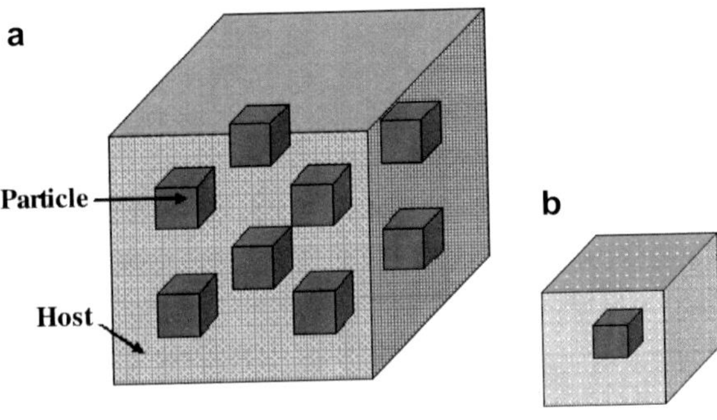

Fig. 10.3 Schematic of **a** Nanoparticle composite and **b** Unit cell

Table 10.1 Material properties calculated using phonon sine function dispersion model

Material	k (W/mK)	$C \times 10^6$ (J/m^{3}K)	v (m/s)	Λ (Å)
Ge	60	0.87	1042	1986
Si	145.6	0.93	1804	2604
GaAs	43.6	0.88	1024	1453
AlAs	86.4	0.88	1246	2364
Bi$_2$Te$_3$	1.1	0.5	212	310
Sb$_2$Te$_3$	0.9	0.53	200	254

Fig. 10.3b is utilized to simulate the effect of a single particle and the surrounding host material within the cell.

The BTE for the phonon intensity in 3-D given by Eq. (10.7) is solved in conjunction with suitable boundary and interface treatments. The material properties used in the computation are evaluated using the phonon sine function dispersion model proposed by Chen. Table 10.1 summarizes the properties of Ge, Si, $GaAs$, $AlAs$, Bi_2Te_3, and Sb_2Te_3 calculated using the dispersion model at 300 K.

In the present case, the interface is treated as totally diffuse and the Diffuse Mismatch Model (DMM) is used. The top and the bottom boundaries of the unit cell are treated as adiabatic surfaces and the remaining boundaries are modeled as periodic, to maintain continuity of heat flow. The adiabatic condition is modeled as a specularly reflecting boundary on which the following condition for intensity is imposed:

$$I\,(\vec{r}_b, \vec{s}, t) = I\,(\vec{r}_b, \vec{s}_r, t) \tag{10.14}$$

where $\vec{s}_r = \vec{s} - 2\,(\vec{s} \cdot \vec{n})\,\vec{n}$, $\vec{n}$ is the outward normal vector, and $\vec{r}_b$ denotes the spatial coordinates of the boundary.

The parameters that have been varied to study their effect on the thermal characteristics of the nanoparticle composite are the particle size, atomic percentage (AP), and the particle aspect ratio (AR). The atomic percentage of the nanoparticle is related to the volumetric percentage, VP, and the lattice constant, a, as

$$AP = \frac{VP}{\left(VP + (1 - VP)\dfrac{a_p}{a_h}\right)}. \tag{10.15}$$

For the cubic nanoparticle composite, $VP = \frac{L_p^3}{L_h^3}$, where L_p and L_h are the lengths of the particle and host respectively. Aspect ratio, AR, is defined as the ratio of nanoparticle length in the z-direction to its length in the x-direction, L_z/L_x.

The phonon intensity obtained from solving the BTE is then used to determine the heat flux, temperature distribution, and thermal conductivities. At nanoscales, the temperature, as such, has no physical meaning except that it is an indicator of the local energy density of the system. The effective temperature is obtained from phonon intensity as

$$T(x, y, z) = \frac{1}{C_h v_h} \sum_{N_\theta} \sum_{N_\phi} I(x, y, z, \theta, \phi) w_\theta w_\phi \qquad (10.16)$$

where w_θ and w_ϕ are the weights associated with the polar and azimuthal directions respectively. The heat fluxes in the x and z directions namely q_x and q_z are related to the intensity through the relations

$$q_x(x, y, z) = \sum_{N_\theta} \sum_{N_\phi} I(x, y, z, \theta, \phi)\, \mu\, w_\theta w_\phi \qquad (10.17)$$

$$q_z(x, y, z) = \sum_{N_\theta} \sum_{N_\phi} I(x, y, z, \theta, \phi)\, \xi\, w_\theta w_\phi. \qquad (10.18)$$

For the non-cubic nanoparticle composite, the effective thermal conductivities in the x-direction, k_x, and in the z-direction, k_z, can be defined as

$$k_x = \frac{Q_x L_x}{L_y L_z (\overline{T}(x=0) - \overline{T}(x=L_h))}; \; k_z = \frac{Q_z L_z}{L_x L_y (\overline{T}(z=0) - \overline{T}(z=L_h))} \qquad (10.19)$$

where Q_x and Q_z are the heat flow in the x- and z-directions respectively. $\overline{T}$ is the area averaged temperature at the boundary. For the cubic nanoparticle composite, $k = k_x = k_y$ due to symmetries in geometry and heat flow in all three coordinate directions.

10.3.1 Temperature Contours

Steady-state heat transport simulations are conducted for a cubic $Sb_2 Te_3$ nanoparticle suspended in a $Bi_2 Te_3$ host material. The nature of phonon transport is characterized by a non-dimensional ratio known as the Knudsen number (Kn), which represents the ratio of the phonon mean free path to the particle size. A large Kn ($Kn >> 1$) indicates a ballistic limit, while a small Kn ($Kn << 1$) suggests a Fourier limit. In Fig. 10.4a, temperature contours ($T - T_{ref}$) are depicted for a Kn value of 10, with heat transport occurring in both the x- and z-directions. Here, the reference temperature T_{ref} is set at 300 K. The temperature contours reveal significant non-linearity in the temperature distribution within both the particle and the host, attributable to pronounced phonon scattering at the interfaces between the two materials. The contours cluster near the interfaces, where interface scattering dominates the internal phonon scattering.

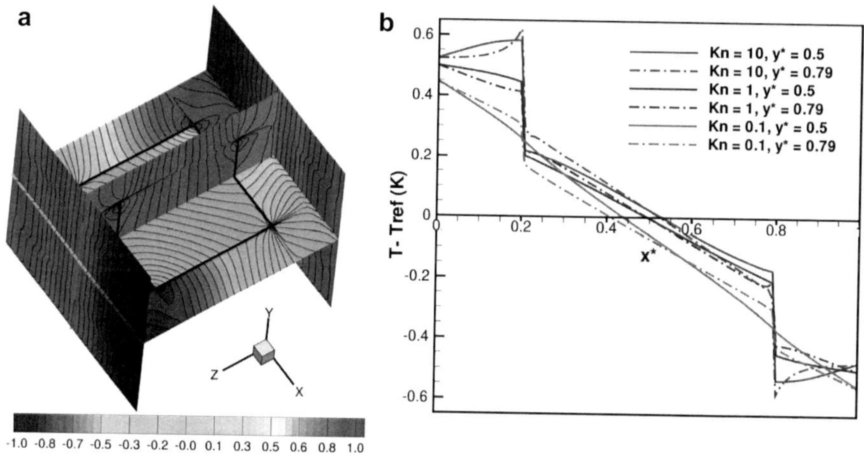

Fig. 10.4 **a** Temperature contours for $Bi_2Te_3 - Sb_2Te_3$ nanoparticle composite for AP = 0.2 and $L_{Sb_2Te_3} = 25.4$ Å and **b** Temperature profiles $(T - T_{ref})$ for AP = 0.2

To further comprehend the effects of scattering from the particle interfaces, temperature profiles are presented at two different y-locations, as shown in Fig. 10.4b. These locations are in the x-y plane at midway $(y* = 0.5)$ and through the top interface $(y* = 0.79)$. Temperature jumps observed at the host-particle interface are attributed to interface thermal resistance, consistent with observations in lower-dimensional nanostructure configurations such as superlattice and nanowire composites.

For a comprehensive comparison of the impact of Kn on temperature, profiles for the ballistic case $(Kn = 10)$ are contrasted with transitional $(Kn = 1)$ and Fourier $(Kn = 0.1)$ cases. In the $Kn = 10$ scenario, the left interface exhibits a higher temperature, while the right interface displays a lower temperature than the boundary temperatures, resulting in a change in the sign of the temperature gradient along the x-direction. This effect is not observed for lower Kn values where ballistic effects are less pronounced. The temperature profiles in the y-z plane, corresponding to heat flow in the z-direction, exhibit a similar trend to the x-direction due to the cubic nature of the nanoparticle and uniform heat fluxes in both directions.

10.3.2 Effective Thermal Conductivity of Nanocomposites

Figure 10.5a illustrates the variation in effective thermal conductivity, denoted as k, for the cubic $Bi_2Te_3 - Sb_2Te_3$ nanoparticle composite with changing particle size at atomic percentages of 0.2 and 0.8. This figure includes thermal conductivity references for the $Bi_2Te_3 - Sb_2Te_3$ superlattice and 2-D $Bi_2Te_3 - Sb_2Te_3$ nanowire. A decrease in thermal conductivity is observed as the particle size diminishes, attributed

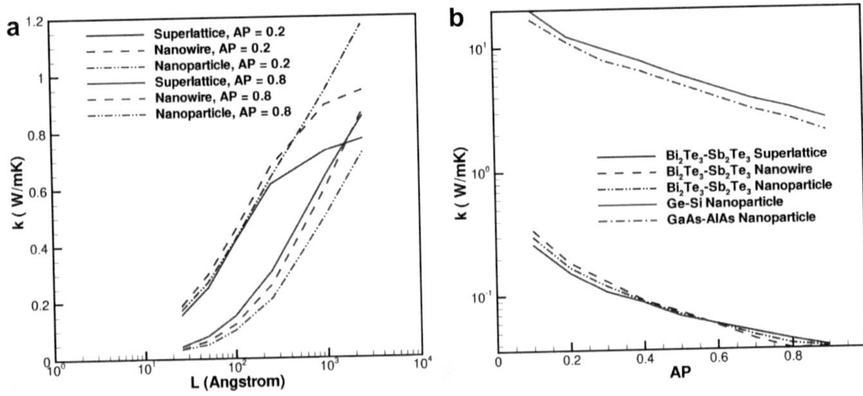

Fig. 10.5 **a** "Size effect" on thermal conductivity of $Bi_2Te_3 - Sb_2Te_3$ nanocomposites. **b** Effect of atomic percentage on thermal conductivity of nanostructures for $Kn = 10$

to increased phonon scattering at the nanoparticle interfaces due to ballistic effects associated with decreasing particle size. This size-dependent thermal conductivity is referred to as the "classical size effect." For $AP = 0.2$, the thermal conductivity values of the cubic nanoparticle composite closely align with those of 1-D superlattice and 2-D nanowire composites. Larger particle sizes lead to thermal conductivities in the nanoparticle, nanowire, and superlattice composites approaching their respective bulk values, with the bulk value of the nanoparticle composite exceeding that of the nanowire and superlattice composites.

However, for $AP = 0.8$, for particle sizes exceeding 100 Å, the thermal conductivity of the nanoparticle composite is lower than that of the nanowire and superlattice composites. For particle sizes below 25 Å, where the phonon wavelength becomes comparable to the nanoparticle dimension, wave effects may play a role in predicting the value of k. Figure 10.5a also indicates that for a given wire size, the thermal conductivity corresponding to $AP = 0.8$ is lower than that of $AP = 0.2$, attributed to an increase in ballistic effects in the host for $AP = 0.8$.

In Fig. 10.5b, the influence of atomic percentage on the thermal conductivity of $Bi_2Te_3 - Sb_2Te_3$, $Ge - Si$, and $GaAs - AlAs$ nanocomposites is presented. The atomic percentage varies from 0.1 to 0.9, while keeping the particle Knudsen number fixed at 10. Increasing the atomic percentage for a given particle size results in a reduction in the host size, leading to heightened phonon interface scattering and, consequently, a decrease in thermal conductivity. The thermal conductivity values for the 3-D $Bi_2Te_3 - Sb_2Te_3$ nanoparticle composite resemble those of the superlattice and nanowire composites. While the $Ge - Si$ and $GaAs - AlAs$ nanoparticle composites exhibit similar atomic percentage effects, their thermal conductivities are approximately two orders of magnitude higher than that of $Bi_2Te_3 - Sb_2Te_3$.

Table 10.2 presents a comparison of thermoelectric properties and the dimensionless figure of merit (ZT) for the $Bi_2Te_3 - Sb_2Te_3$ superlattice and nanocomposite.

Table 10.2 Thermoelectric properties of $Bi_2Te_3 - Sb_2Te_3$ superlattice and nanocomposite

Material	Method	AP	p	k (W/mK)	$S^2\sigma$ (10^{-3} W/mK^2)	ZT
$Bi_2Te_3 - Sb_2Te_3$ SL	Experiment [97]	0.5	–	0.22	1.76	2.4
$Bi_2Te_3 - Sb_2Te_3$ SL	Simulation	0.5	0.9	0.28	1.76	1.89
$Bi_2Te_3 - Sb_2Te_3$ NC	Simulation	0.5	0.9	0.26	1.76	2.03
$Bi_2Te_3 - Sb_2Te_3$ NC	Simulation	0.1–0.9	0–1	0.034–0.74	1.76	0.71–15.5

The calculated thermal conductivity (k) values for the nanocomposite and superlattice, corresponding to atomic percentages (AP) of 0.5 and specularity parameter (p) of 0.9, are 0.26 and 0.28, respectively. The choice of $p = 0.9$ is based on achieving the best match between simulation and experimental results for the superlattice.

For the nanocomposite, k values range from 0.034 to 0.74, depending on the atomic percentage and interface treatment. Although electron properties like the Seebeck coefficient and electrical conductivity may vary for the nanocomposite configuration, for comparison purposes, the power factor of the superlattice is used to estimate ZT for the nanocomposite. Assuming the power factor $S^2\sigma$ is equal to that of the superlattice, ZT values in the range of 0.71 to 15.5 are obtained for the nanocomposite. This underscores the significance of utilizing $Bi_2Te_3 - Sb_2Te_3$ nanostructures in efficient thermoelectric applications.

10.3.3 Comparison of BTE with Fourier Results

To underscore the limitations of the Fourier model when applied to heat transport in nanoparticle composites, a comparison is drawn between the Boltzmann Transport Equation (BTE) and Fourier solutions, as depicted in Fig. 10.6. Figure 10.6a and c illustrates the relative percentage differences between BTE and Fourier temperature predictions for the non-cubic $Bi_2Te_3 - Sb_2Te_3$ nanoparticle composite in the x-y and x-z planes, respectively. The relative percentage difference is calculated as $(T_{BTE} - T_{Fourier}) \times 100/T_{BTE}$. Positive values indicate under-prediction by the Fourier model. Notably, regions near the interface of the two materials, where temperature maxima and minima occur, exhibit the most significant differences between the two models. The Fourier model under-predicts regions of maximum temperature by up to 50%, while over-predicting regions of minimum temperature by 30%. These findings align with results from nanoscale Silicon-on-Insulator (SOI) structures, where the Fourier solution significantly underestimates peak phonon temperatures or "hot spots."

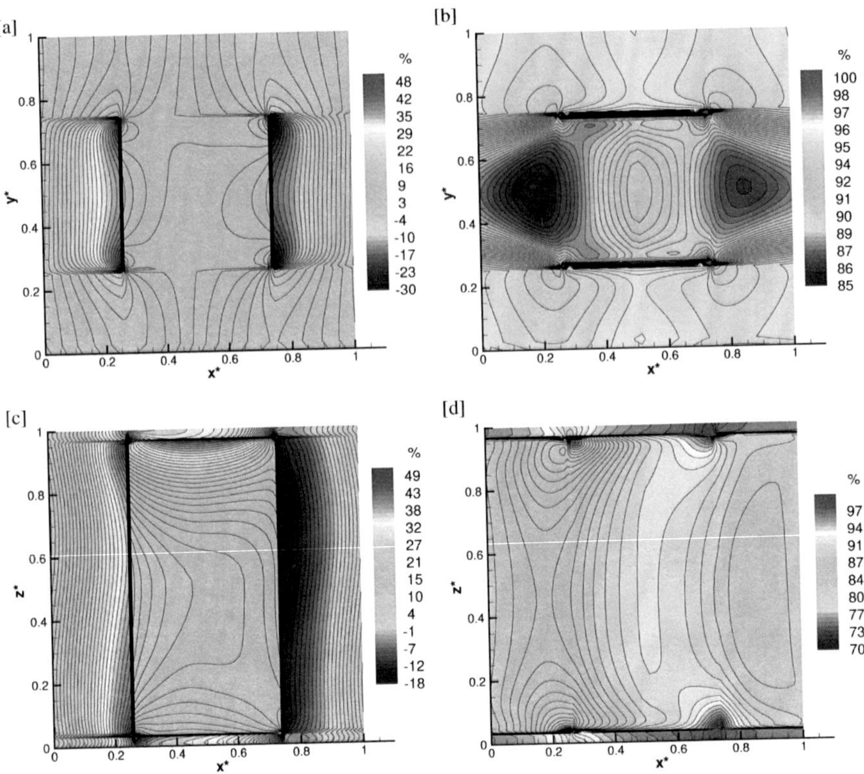

Fig. 10.6 Percentage difference between BTE and Fourier simulations as observed in (**a**), (**c**) temperature contours and (**b**), (**d**) heat flux contours for a cubic and non-cubic $Bi_2Te_3 - Sb_2Te_3$ nanoparticle composite $Kn = 10$

In Fig. 10.6b and d, the percentage difference in x-direction heat fluxes between the BTE and Fourier models is shown for the two planes. The Fourier model consistently underestimates the heat flux throughout the nanocomposite by as much as 100%. In the x-y plane, the percentage difference ranges between 85% and 100%, while in the x-z plane, it varies between 70% and 100%.

This chapter has explored energy transport in nanostructures based on the Boltzmann Transport Equation using computational methods. The ability to predict thermal behavior at nanoscales remains crucial for advancements in thermoelectrics and micro/nano electronics.

Exercise Problems

1. Write a program to solve the 1-D Boltzmann Transport Equation describing phonon transport in thin films as described in Sect. 5.2. Plot the temperature profiles and effective thermal conductivity of the thin film as a function of Knudsen Number (classical size effect).
2. Write a program to solve the 1-D Boltzmann Transport Equation describing rarified conduction in a stationary gas. Plot the temperature profiles and effective thermal conductivity of the gas slab as a function of Knudsen Number.

References

1. Adachi, S. (ed.): Properties of Aluminium Gallium Arsenide. INSPEC, London (1993)
2. Agarwal, R.K., Yun, K.Y., Balakrishnan, R.: Beyond Navier-Stokes: Burnett equations for flows in the continuum-transition regime. Phys. Fluids **13**, 3061–3085 (2001)
3. Agrawal, A., Prabhu, S.V.: Survey on measurement of tangential momentum accommodation coefficient. J. Vac. Sci. Technol. A **26**, 634–645 (2008)
4. Ameel, T.A., Wang, X., Barron, R.F., Warrington, R.O.J.: Laminar forced convection in a circular tube with constant heat flux and slip flow. Microscale Thermophys. Eng. **1**, 303–320 (1997)
5. Azer, N., Abis, L., Soliman, H.: Local heat transfer coefficients during annular flow condensation. ASHRAE Trans. **78**, 135–143 (1972)
6. Bae, S., Maulbetsch, J.S., Rohsenow, W.M.: Refrigerant Forced-Convection Condensation Inside Horizontal Tubes. MIT Heat Transfer Laboratory, Cambridge, MA
7. Baker, O.: Simultaneous flow of oil and gas. Oil Gas J. **53**, 185–190 (1954)
8. Bandhauer, T.M., Agarwal, A., Garimella, S.: Measurement and modeling of condensation heat transfer coefficients in circular microchannels. J. Heat Transf. **128**, 1050–1059 (2006)
9. Barnea, D., Luninski, Y., Taitel, Y.: Flow pattern in horizontal and vertical two phase flow in small diameter pipes. Can. J. Chem. Eng. **61**, 617–620 (1983)
10. Baroczy, C.: Systematic correlation for two-phase pressure drop. Chem. Eng. Prog. Symp. Ser **62**, 232–249 (1966)
11. Boyko, L.D., Kruzhilin, G.N.: Heat transfer and hydraulic resistance during condensation of steam in a horizontal tube and in a bundle of tubes. Int. J. Heat Mass Transf. **10**, 361–373 (1967)
12. Brauner, N., Maron, D.M.: Identification of the range of 'small diameters' conduits, regarding two-phase flow pattern transitions. Int. Commun. Heat Mass Transf. **19**, 29–39 (1992)
13. Breber, G., Palen, J.W., Taborek, J.: Prediction of horizontal tubeside condensation of pure components using flow regime criteria. J. Heat Transf. **102**, 471–476 (1980)
14. Butterworth, D.: A comparison of some void-fraction relationships for co-current gas-liquid flow. Int. J. Multiph. Flow **1**, 845–850 (1975)
15. Carey, V.: Liquid-Vapor Phase-Change Phenomena. Taylor and Francis Group, New York, NY
16. Carnavos, T.: Heat transfer performance of internally finned tubes in turbulent flow. Heat Transf. Eng. **1**, 32–37 (1980)
17. Chien, A.: Modelling Energy Transport in Nanostructures. Published doctoral thesis, State University of New York at Buffalo (2009)

© The Author(s) 2025

A. Pattamatta and S. K. Das, *Fundamentals of Nano- and Microscale Heat Transport,*
https://doi.org/10.1007/978-3-031-89613-2

18. Chieh, H., Yao, D.J., Hsu, C.T.: Measurement and evaluation of the interfacial thermal resistance between a metal and a dielectric. Appl. Phys. Lett. **93**, (2008)
19. Ashcroft, N.W., Mermin, N.D.: Solid State Physics. Saunders College, Philadelphia (1976)
20. Bannov, N., Aristov, V., Storcio, M.A.: Electron relaxation time due to the deformation-potential. interaction of electron with confined acoustic phonons in a free-standing quantum well. Phys. Rev. **51**, 9930–9942 (1995)
21. De Broglie, L.: Recherchus sur la Theorie des Quanta. Ann. Phys. **3**, 22 (1925)
22. Callen, H.B.: Thermodynamics and an Introduction to Thermostatics. Wiley, New York
23. Casimir, H.B.G.: Note on the conduction of heat in crystals. Physica **5**, 495–500 (1938)
24. Chen, G.: Size and interface effects on thermal conductivity of superlattices and periodic thin-film structures. J. Heat Transf. **119**, 220–229 (1997)
25. Chen, G.: Heat transfer in micro and nanoscale photonic devices. Annu. Rev. Heat Transf. **7**, 1–58 (1996)
26. Chen, G.: Phonon Heat Conduction in Low-Dimensional Structures. In: Semiconductors and Semimetals. Recent Trends in Thermoelectric Materials Research-3, vol. 71, pp. 203–259. Academic Press, San Diego (2001)
27. Chen, G., Borca-Tasciuc, D., Yang, R.G.: Nanoscale heat transfer. Encycl. Nanosci. Nanotechnol. **17**, 429–459 (2004)
28. Davisson, C., Germer, L.H.: Diffraction of electrons by a crystal of nickel. Phys. Rev. **30**, 705–740 (1927)
29. Das, S.K., Putra, N., Thiesen, P., Roetzel, W.: Temperature dependence of thermal conductivity enhancement for nanofluids. J. Heat Transf. **125**, 567–574 (2003)
30. Das, S.K., Putra, N., Roetzel, W.: Pool boiling characteristics of nano-iñĆuids. Int. J. Heat Mass Transf. **46**, 851–862 (2003)
31. Das, S.K, Choi, S.U.S., Yu, W., Pradeep, T.: Nanofluids: Science and Technology. John Wiley & Sons (2008)
32. Dhar, P., Gupta, S.S., Chakraborthy, S., Pattmatta, A., Das, S.K.: The role of percolation and sheet dynamics during heat conduction in poly-dispersed graphene nanofluids. Appl. Phys. Lett. **102**, 163114 (2013)
33. Dhar, P., Ansari, M.H.D., Gupta, S.S., Siva, M.V., Pradeep, T., Pattmatta, A., Das, S.K.: Percolation network dynamicity and sheet dynamics governed viscous behavior of poly-dispersed Graphene nano-sheet suspensions. J. Nanoparticle Res. **15**, 2095 (2013)
34. Einstein, A.: On a heuristic point of view concerning the production and transformation of light. Ann. Phys. **17**, 132–148 (1905)
35. Einstein, A.: On the theory of light production and light absorption. Ann. Phys. **20**, 199–206 (1906)
36. English, N.J., Kandlikar, S.G.: An experimental investigation into the effect of surfactants on air-water two-phase flow in minichannels. In: Proceedings of the third international conference on microchannels and minichannels, ASME Paper No. ICMM 2005-75110, June 13–15, Toronto, Canada, p 8 (2005)
37. English, N.J., Kandlikar, S.G.: An experimental investigation into the effect of surfactants on air-water two-phase flow in minichannels. Heat Transfer Eng. **27**(4), 99–109 (2006)
38. Esaki, L., Tsu, R.: Superlattice and negative differential conductivity in semi-conductors. IBM J. Res. Dev. **14**, 61–65 (1970)
39. Feynman, R.P.: 1959, there is a plenty of room at the bottom, talk at annual meeting of American physical society, December 26. J. Microelectromechanical Syst. **1**, 60–66 (1992)
40. Feynman, R.P.: 1983, infinitesimal machinery, talk at jet propulsion laboratory, February 23. J. Microelectromechanical Syst. **2**, 4–14 (1993)
41. Feynman, R.: Lectures on Physics, vol 3, Chapters 1–3. Addison-Wesley, Reading, MA (1965)
42. Fleming, J.G., Lin, S.Y., El-Kady, I., Biswas, R., Ho, K.M.: All-metallic three dimensional photonic crystals with a large infrared bandgap. Nature **417**, 52–55 (2002)
43. Fukui, S., Kaneko, R.: Analysis of ultra-thin gas film lubrication based on linearized Boltzmann equation: first report-derivation of a generalized lubrication equation including thermal creep flow. ASME J. Tribol. **110**, 253–263 (1988)

44. Ghatak, K., Biswas, S.: Influence of quantum confinement on the heat capacity of graphite. Fizika A **3**, 7–24 (1994)
45. Goldsmid, H.J.: Thermoelectric Refrigeration. Plenum Press, New York (1964)
46. Griffiths, D.J.: Introduction to Quantum Mechanics. Prentice Hall, Englewood Cliffs, NJ (1994)
47. Harirchian, T., Garimella, S.V.: Effects of channel dimension, heat flux, and mass flux on flow boiling regimes in microchannels. Int. J. Multiph. Flow **35**, 349–362 (2009)
48. Harman, T.C., Taylor, P.J., Walsh, M.P., LaForge, B.E.: Quantum dot superlattice thermoelectric materials and devices. Science **297**, 2229–2232 (2002)
49. Heavens, O.S.: Optical Properties of Thin Solid Films. Dover Publications, New York (1965)
50. Holland, M.G.: Phonon scattering in semiconductors from thermal conductivity studies. Phys. Rev. **132**, 2461–2471 (1964)
51. Heisenberg, W.: Uber Quantentheoretische Umdeutung Kinematischer und Mechanischer Beziehungen. Zeitschrift fur Physik **33**, 879–893 (1925)
52. Iijima, S.: Helical microtubules of graphitic carbon. Nature **354**, 56–58 (1991)
53. Imry, Y., Landauer, R.: Conductance viewed as transmission. Rev. Mod. Phys. **71**, S306–S312 (1999)
54. Incropera, F.P., DeWitt, D.P.: Fundamentals of Heat and Mass Transfer. 6th edition, Wiley (2002)
55. Ju, Y.S., Goodson, K.E.: Phonon scattering in silicon films with thickness of order 100 nm. Appl. Phys. Lett. **74**, 3005–3007 (1999)
56. Kakak, S., Shah, R.K., Aung, W.: Hand Book of Single-Phase Convective Heat Transfer. John Wiley and Sons Inc, New York, NY (1987)
57. Kandlikar, S.G.: A general correlation for two-phase flow boiling heat transfer inside horizontal and vertical tubes. J. Heat Transf. **112**, 219–228 (1990)
58. Kandlikar, S.G.: A model for flow boiling heat transfer in augmented tubes and compact evaporators. J. Heat Transf. **113**, 966–972 (1991)
59. Kandlikar, S.G., Balasubramanian, P.: An extension of the flow boiling correlation to transition, laminar and deep laminar flows in minichannels and microchannels. Heat Transf. Eng. **25**, 86–93 (2004)
60. Kandlikar, S.G., Schmitt, D., Carrano, A.L., Taylor, J.B.: Characterization of surface roughness effects on pressure drop in single-phase flow in minichannels. Phys. Fluids **17**, (2005)
61. Kandlikar, S.G., Garimella, S., Li, D., Colin, S., King, M.R.: Heat Transfer and Fluid Flow in Minichannels and Microchannels. Netherlands; San Diego, CA, Oxford, Elsevier, UK, Amsterdam (2005)
62. Kandlikar, S.G., Steinke, M.E.: Predicting Heat Transfer During Flow Boiling in Minichannels and Microchannels. ASHRAE Trans., **109**, Part 1 (2003)
63. Karniadakis, G.E., Beskok, A.: Microflows: Fundamentals and Simulations. Springer, New York, NY (2002)
64. Kirby, B.J.: Micro and Nanoscale Fluid Mechanics: Transport in Micro Fluidic Devices. Cambridge University Press (2010)
65. Kittel, C., Kroemer, H.: Thermal Physics. Wiley, New York
66. Kubo, R., Toda, M., Hashitsume, N.: Statistical Physics 2, 2nd edn. Springer, Berlin (1998)
67. Landau, L.D.: The theory of super fluidity of helium 2. J. Phys. (Mosc.) **5**, 71–90 (1941)
68. Landau, L.D., Lifshitz, E.M.: Quantum Mechanics. Pergamon, Oxford (1977)
69. Lasjaunias, J.C., Biljakovic, K., Monceau, P., Sauvajol, J.L.: Low-energy vibrational excitations in carbon nanotubes studied by heat capacity. Nanotechnology **14**, 998–1003 (2003)
70. Lee, S.M., Cahill, D.G.: Heat transport in thin dielectric films. J. Appl. Phys. **81**, 2590–2595 (1997)
71. Lee, S., Choi, S.U.S., Li, S., Eastman, J.A.: Measuring thermal conductivity of fluids containing oxide nanoparticles. J. Heat Transf. **121**, 280–289 (1999)
72. Li, Q., He, Y., Tang, G., Tao, W.: Lattice Boltzmann modeling of microchannel flows in the transition flow regime. Microfluid. Nanofluid **10**, 607–618 (2011)

73. Lim, C.Y., Shu, C., Niu, X.D., Chew, Y.T.: Application of lattice Boltzmann method to simulate microchannel flows. Phys. Fluids **14**, 2299–2308 (2002)

74. Maiga, S.E., Palm, S.J., Nguyen, C.T., Roy, G., Galanis, N.: Heat transfer enhancement by using NanoïñĆuids in forced convection flows. Int. J. Heat Fluid Flow **26**, 530–546 (2005)

75. Mishima, Hibiki: Some Characteristics of Air-Water Two-Phase Flow in Small Diameter Vertical Tubes. Int J. Multiph. Flow **22**, 703–712 (1996)

76. Morales, A.M., Lieber, C.M.: A laser ablation method for the synthesis of crystalline semiconductor nanowires. Science **279**, 208–211 (1998)

77. Narayanaswamy, A., Chen, G.: Surface modes for near-field thermophotovoltaics. Appl. Phys. Lett. **82**, 3544–3546 (2003)

78. Pak, B., Cho, Y.I.: Hydrodynamic and heat transfer study of dispersed fluids with submicron metallic oxide particle. Exp. Heat Transf. **11**, 151–170 (1998)

79. Pattamatta, A.: Modelling energy transport in nanostructures. Published doctoral thesis, State University of New York at Buffalo (2009)

80. Putra, N., Roetzel, W., Das, S.K.: Natural convection of nanofluids. Heat Mass Transf. **39**, 775–784 (2003)

81. Phillips, R.J.: Forced convection, liquid cooled, microchannel heat sinks, (MS thesis), Department of Mechanical Engineering, Massachusetts Institute of Technology, Cambridge, MA (1987)

82. Polder, D., Van Hove, M.: Theory of radiative heat transfer between closely spaced bodies. Phys. Rev. B **4**, 3303–3314 (1971)

83. Prasher, R.S., Phelan, P.E.: Size effects on the thermodynamic properties of thin solid films. J. Heat Transf. **120**, 1078–1086 (1998)

84. Rayleigh, J.W.S.: The Theory of Sound. Dover, New York

85. Ren, Z.F., Huang, Z.P., Xu, J.W., Wang, J.H., Bush, P., Siegal, M.P., Provencio, P.N.: Synthesis of large arrays of well-aligned carbon nanotubes on glass. Science **282**, 1105–1107 (1998)

86. Schmitt, D.J., Kandlikar, S.G.: Effects of repeating microstructures on pressure drop in rectangular minichannels, Paper No. ICMM 2005-75111. In: Third international conference on microchannels and minichannels, Toronto, Canada, 13–15 June. ASME (2005)

87. Schrodinger, E.: Quantisation as a problem of characteristic values. Annalen der physik **79**, 361–376 (1926)

88. Shur, M.: Physics of Semiconductor Devices. Prentice Hall, Englewood Cliffs, NJ (1990)

89. Srinikaeth, T.: Thermal conductivity and thermal interface resistance measurement of SiO_2 thin films using 3ω method. B. Tech thesis, Indian Institute of Technology Madras (2016)

90. Steinke, M.E., Kandlikar, S.G.: Single-phase liquid friction factors in microchannels Paper No. ICMM2005-75112. In: Third international conference on microchannels and minichannels, Toronto, Canada, 13–15 June (2005a)

91. Stillinger, F.H., Weber, T.: Computational simulation of local order in condensed phases of silicon. Phys. Rev. B **31**, 5262–5271 (1985)

92. Sze, S.M.: Physics of Semiconductor Devices. Wiley, New York (1981)

93. Tien, C.L., Cunnington, G.R.: Cryogenic insulation heat transfer. Adv. Heat Transf. **9**, 349–417 (1973)

94. Tsu, R., Esaki, L.: Tunneling in a finite superlattice. Appl. Phys. Lett. **22**, 562–564 (1973)

95. Vassallo, P., Kumar, R., D'Amico, S.: Pool boiling heat transfer experiments in silica-water nano-fluids. Int. J. Heat Mass Transf. **47**, 407–411 (2004)

96. Weisbuch, C., Vinter, B.: Computational simulation of local order in condensed phases of silicon. In: Quantum Semiconductor Structures. Academic Press, Boston

97. Xuan, Y., Li, Q.: Investigation on convective heat transfer and flow features of nanoïñĆuids. J. Heat Transf. **125**, 151–155 (2003)

98. Yang, B., Chen, G.: Lattice dynamics study of phonon heat conduction in quantum wells. Phys. Low Dimens. Struct. J. **5**(6), 37–48 (2000)

99. Yan, Y., Lin, T.: Evaporative heat transfer and pressure drop of refrigerant R-134a in a small pipe. Int. J. Heat Mass Transf. **42**, 4183–4194 (1998)

100. Yariv, A.: Quantum Electronics. Wiley, New York

101. Ziman, J.M.: Chapter 3. Clarendon Press, Oxford (1960)
102. Venkatasubramanian, R.: Lattice thermal conductivity reduction and phonon localization like behavior in superlattice structures. Phys. Rev. B **61**, 3091–3097 (2000)
103. Pattamatta, A., Madnia, C.K.: Modeling thermal transport in nanoparticle composites. In: 40th AIAA Thermophysics Conference (2008)
104. Pattamatta, A., Madnia, C.K.: Modeling heat transfer in $Bi_2Te_3 - Sb_2Te_3$ nanostructures. Int. J. Heat Mass Transf. **52**, 860–869 (2009)
105. Yang, R.: Nanoscale heat conduction with applications in nanoelectronics and thermoelectrics. Thesis (Ph.D.)–Massachusetts Institute of Technology, Department of Mechanical Engineering (2006)